# Software as Hermeneutics

Luca M. Possati

# Software as Hermeneutics

## A Philosophical and Historical Study

palgrave
macmillan

Luca M. Possati
Institute of Philosophy
University of Porto
Porto, Portugal

ISBN 978-3-030-63609-8 ISBN 978-3-030-63610-4 (eBook)
https://doi.org/10.1007/978-3-030-63610-4



This Palgrave Macmillan imprint is published by the registered company Springer Nature Switzerland AG.
The registered company address is: Gewerbestrasse 11, 6330 Cham, Switzerland

# Contents

# List of Figures

# List of Tables

# 1

# Introduction: The Paradoxical Onion

Human beings are the only animals capable of inventing, constructing, and manipulating signs. This ability gives them access to reflective consciousness, culture, history, and science. Today as never before in the history of humanity, signs play a fundamental role in the construction and management of human society overall. Thanks to digital technology, signs have undergone a radical transformation: they are not just instruments, simple "representatives of" as symbol, icon, or index following Peirce's categories. They have acquired an autonomous life and become social agents in all respects, able to perform jobs, interact with humans, and plan events without human interaction. This radical transformation of signs has a name: *software*. The present research aims to show that the growth of software is the last outcome of a much older history, dating back to the seventeenth and eighteenth centuries, that has its central point in the formation of modern mathematical writing. This does not mean identifying the history of software with that of computation. I argue that if we want to understand a phenomenon as pervasive as software, we must first study the essence of a simple gesture, which we perform every day: that of writing. The "digital revolution" (in the critical sense of this expression used in Vial 2013 and Salanskis 2011) is a revolution of writing.



L. M. Possati, *Software as Hermeneutics*,
https://doi.org/10.1007/978-3-030-63610-4_1

The book you have in your hands is a book on philosophy and the history of philosophy and science. It is a book that, through historical investigations, tries to develop some philosophical theses on the nature of this mysterious object that is software. One of these theses is that the digital object (the icons on the screen of my laptop, the apps on my smartphone, the cloud I use to store my documents, the newspaper I read on my tablet, the music I listen to on my phone, or my Instagram account) is a written object; its ultimate essence is writing. How can a physical object be made (constructed, made possible) by writing? If we analyze an apple using a microscope, we see cells and biological structures, and beyond these structures, the quantum reality. If we analyze the digital object, *we see a writing in it*, that is, the code, software. My argument is that this writing is the condition of possibility of the digital object, but it is a condition that never shows itself, that we never see directly, like the biological and atomic structures of the apple. As Elliott (2018, 31) claims, this is a general characteristic of digital technology:

> Digital life inaugurates a transformation in the nature of invisibility—operationalized through supercomputers, big data, and artificial intelligence—and the changing relation between the visible, the hidden and power. My argument is that the rise of systems of digital technology in the late twentieth and early twenty-first centuries has created a new form of invisibility which is linked to the characteristics of software code, computer algorithms and AI protocols and to its modes of information processing.

I do not have to know programming languages to be able to use a program on my laptop, just as I do not have to know chemistry to enjoy the taste of an apple. The computer, as well as the smartphone or iPad, has become a "black box": we use graphical interfaces and images to avoid interacting with something that remains hidden—namely, a deep writing. The interface makes the software a perceptible object, but at the same time it creates a distance between us and software. Software itself is a noumenon, that is, a phenomenon without phenomenality—a phenomenon that does not manifest itself and that we cannot directly access by perception.[1]

One of the main purposes of this book is to justify these theses by giving as rigorous a definition of software as possible and from the point of view of continental philosophy. This book wants to show that continental philosophy can give us a new understanding of digital technology. This does not mean undermining the contribution of analytic philosophy or other disciplines; instead, it means arguing that continental philosophy can contribute positively to this debate.

I argue that software is a very complex object, composed of multiple layers (technical documents, HHL, compiler, assembler, machine code, etc.), and that each of these layers corresponds to a process of writing and re-writing, and so to several hermeneutic processes. Software is the only writing that is not made to be read, because "for a computer, to read is to write elsewhere" (Chun 2013, 91). "Software is a special kind of text, and the production of software is a special kind of writing" (Sack 2019, 35). "Software's specificity as a technology resides precisely in the relationship it holds with writing, or, to be more precise, with historically specific forms of writing" (Frabetti 2015, 68). The second fundamental thesis of this research is that writing is already a form of interpretation, what I call a "material hermeneutics." My approach, therefore, differs from that of Frabetti, Chun, and Sack because it connects writing to hermeneutics. Hence, it is based on the connection between hermeneutics and post-phenomenology.

I want to clarify two points now: why a philosophical, and above all hermeneutic, approach to software is useful today, and how I connect hermeneutics and post-phenomenology.

I argue that analyzing software through the conceptual tools of continental philosophy, and especially hermeneutics, gives us an innovative point of view on software—on its design, construction, and even its use. In other words, a hermeneutic perspective on software gives us a global understanding of software that can also be useful to the programmer in their job. Software hermeneutics therefore intends to be a support of software engineering and data analytics.

This research is based on a certain conception of the philosophy of technology and AI. Digital technology is an extraordinary source of philosophical questions, that is, questions which are "in principle *open*, *ultimate*, *closed under further questioning*, and possibly *constrained* by empirical and logico-mathematical resources, which require *noetic resources* to be answered" (Floridi 2019, 19).[2] Jean-Claude Beaune in *L'automate et ses mobiles* (1980, 10) holds that technical devices have always been "philosophical machines" that are capable of producing not only new types of entities, but also new ways of perceiving and interpreting reality. In *L'être et l'écran* (2013), Stéphane Vial changes the expression—he prefers to use "ontophanic matrixes"—but the concept does not change: technology generates "a priori structures of perception, historically dated and culturally variable." Digital devices, "like all technical devices in general, are materialized theories of reality, or reified philosophies of reality" (Vial 2013, 19–2).

Technology is an extraordinary source of philosophical questions because in the relationship with the machine—proportionally to the intensity of this relationship—humans put not only their technical skills and knowledge, but also their world, and this in a phenomenological sense, that is, their relationship with the "things themselves" and with their meaning. This is what Günther Gotthard says, in his fundamental book *Das Bewusstsein der Maschinen: eine Metaphysik der Kybernetik* (1957). Günther claims that digital technologies and AI have triggered an unexpected metaphysical revolution in the history of humanity. In these machines that are neither mere objects nor complete subjects, humans project their own subjectivity, reflection, and worldview. Günther's thesis is that in order to truly understand the cybernetic ego, we must think of a new form of trivalent logic that does not call into question the I/non-ego polarity. This is the central problem of cybernetics: "The question is not what life, consciousness, or self-reflection ultimately is, but: *can we repeat in machines the behavioral traits of all those self-reflective systems that our universe has produced in its natural evolution?*" (Günther 1962, emphasis added). Therefore, the human being is in the machine as much as the machine is in the human being. In *Du mode d'existence des objets techniques* (1958), Gilbert Simondon argues the same thesis: "What resides in machines is human reality, the reality of the human gesture fixed and crystallized in structures that work" (Simondon 2012, 12). This is also the central thesis of Bruno Latour's Actor-Network Theory: our relationships with the objects that surround us are as important as those with other subjects (Latour 2005).

The connection between technology and human reality is even more evident today. The technical system (Gille 1978) has become a branched and complex apparatus, characterized by the unity of four components: science, engineering, industry, and design. This apparatus, developed over the last fifty years, is able to influence every social and political choice, at macro and micro levels. And I think that we should perhaps add a fifth component: the infrastructural system, which connects countries and cities. Nowadays countries or cities count not so much for the wealth or the territorial extent they have, but for how much they are connected with the rest of the world. This trend is revolutionizing traditional political geopolitical equilibria (Parag Khanna 2016; more on it in section "Software Is Eating the World") as well as social dimensions.

Hence, technology produces philosophical problems because humans project their relationship with the world into it. Does this mean that every human relationship with the world is technologically mediated? Can we really affirm that "being is *poiesis*, or an anthropotechnical construction" and that "we have always lived an augmented reality" (Vial 2013, 25, 105)? Is there a "fundamental technical nature of being" (Huyghe 2002, 9)? Such a radical thesis can be dangerous because it can produce an oversimplified image of reality or fuel a technological determinism. Often, philosophers have been blinded by the technique—and not only by that. The same distinction between what is technology and what is not (the "non-technological Garden" in Ihde 1991) is ambiguous: is fire technology? Is the kitchen a technology or an art? Not everything can be reduced to technology; human intentionality is not always technologically mediated. There are dimensions of human existence that do not involve the mediation of techniques and cannot be reduced to data—that is, to strings of 1 and 0. Therefore, technique is a fundamental instrument of interpretation of reality, but it is not the only one. Humans give meaning to reality not just by creating technical objects. Technology comes up, and therefore becomes necessary, when perceptive modalities, needs, and activities of a certain type come up. Then, technology produces new perceptive modalities, needs, activities, and concepts.

* * *

Why post-phenomenology and hermeneutics to look at software, rather than, for example, the actor-network theory, new materialism, deconstruction, or many other possible approaches? If the reason is that, in this way, we can achieve a better understanding of the ontology of software, then the question becomes: Why should we be interested in producing an ontology of software at all? I answer this question in the first chapter of this book. Software introduces a new way of being that forces us to rethink traditional ontological categories.

I respond to the ontological challenge of software through a phenomenological and hermeneutic method. Why this choice? For two reasons. The first is that software is an artifact, that is, something human-made for humans. Therefore, it is something that, like any artifact, is deeply rooted in human experience and its needs. Like any technique, software is also a mediation between the human being and the world. In this sense, my method is post-phenomenological.

But what do I mean here by "post-phenomenological"? Of course, among my points of reference are the works of Ihde and Verbeek, which are the classics of post-phenomenology. This book is also intended to be an introduction to some essential themes of post-phenomenological reflection. From this point of view, the audience to which it is addressed is very broad: from students in computer science, media studies, and cultural studies, to philosophers interested in technology and up to and including researchers.

However, by "post-phenomenology" I also mean Vial's method, which is more inspired by Bachelard. According to Bachelard, every kind of technique produces phenomena, that is, new forms of perception that are historically determined (Bachelard 1933, 1934, 1938). "Every use of the technique is always a phenomenotechnical exercise [*exercice phénoménotechnique*] even if the level of phenomenality is weak or invisible" (Vial 2013, 104). The digital revolution is first a phenomenological revolution *in the subjects*; it reconfigures the ability of humans to perceive the world. By doing this, technology interprets the world. This means that it configures and reconfigures our understanding of being-in-the-world (Heidegger). The technical phenomenon confers an order and a sense on our experience, but it does so only in order to re-signify our experience, that is, to give it new meaning. It is true that "*all our objects* prepare us for the world and participate in a phenomenological way, at different levels, in the ontophanic process of reality" (Vial 2013, 105, emphasis added), but what characterizes *technical objects* is the interpretation that is added to the ontophanic process. In other words, technology produces new phenomena (new ways of experiencing the world) and these new phenomena involve a new interpretation of the world and of the relationship between humans and the world. What interests me in Vial is this concept of ontophany—technology as active producer of phenomena.

I argue that post-phenomenology and Vial's method can be enormously strengthened by grafting hermeneutics onto them. I introduce this "grafting" in the second chapter of this book. Here I analyze the real contribution that continental philosophy can make to the study of software and digital technologies. My argument is that the software can be considered as a text, in the sense of Paul Ricoeur. Software is a text, so it involves a hermeneutic process divided into three phases: prefiguration, configuration, and refiguration. A hermeneutic understanding of software allows us to explain those aspects of software that escape a strictly technical definition, such as the relationship with the user, the human being, and the social and cultural transformations that software produces, the same way that any text does.

The third chapter completes the phenomenological and hermeneutical approach of the second. Here software is analyzed as a technical phenomenon rooted in a long history, that of writing and literacy, which has known different phases and revolutions. Starting from classic historical studies such as Havelock (1963), Leroi-Gourhan (1964), Goody (1977), and Ong (1982), the book shows how writing restructures thought; writing is the first fundamental technology that has transformed the human mind more than any other invention. Indeed, a fundamental fact takes place in writing: the language becomes autonomous from its author and from any social and historical context. This autonomy of the linguistic dimension is a crucial factor in the revolution at the origins of modern mathematics. Viète, Descartes, and Leibniz created a new form of writing that allowed for enormous progress in algebra and calculation. In this kind of mathematics, writing—even in the form of the diagram—has a heuristic function (Serfati 2005).

What does this story say from a post-phenomenological and hermeneutic point of view? As mentioned above, one of the crucial theses of this book is that software is essentially writing. Precisely for this reason, it is intrinsically hermeneutic. Writing is not just a technical fact; it is instead a hermeneutic act, a material hermeneutic.

## A Dominant Language

> One definition of a modern computer is that it is a system: an arrangement of hardware and software in hierarchical layers. Those who work with the system at one level do not see or care about what is happening at other levels. The highest levels are made up of "software"—by definition things that have no tangible form but are best described as method of organization. (Ceruzzi 2003, 4)

We can distinguish four huge epochs in the development of the information revolution:

1. The first commercialization of the computer between the 1940s and 1950s, with the birth of the UNIVAC project, the first example of a computer entirely dedicated to commercial use. The real turning point of this phase is the invention of the stored program principle developed by Eckert and Mauchly and then theorized by von Neumann (Ceruzzi 2003, 23). A large number of UNIVAC's users were private companies and not the government or the military. In this phase, the first forms of software are developed in the United States, with the electromechanical Harvard Mark I, and then the first types of compiler by J. H. Laning and N. Zierler at MIT. The programming language FORTRAN was introduced by IBM for the 704 computer in early 1957 (Ceruzzi 2003, 89–90).
2. The birth and development of minicomputers in the 1960s and 1970s. The minicomputer opened up a whole area of new applications (Ceruzzi 2003, 124). With it, the computer became a personal interactive device. An example of a microcomputer is the MITS Altair

8800. The use of transistors "made these machines compact, allowing them to be installed in the same rooms that previously housed the array of tabulations and card punches that typically handled a company's data-processing needs" (Ceruzzi 2003, 75).

3. The birth and development of the personal computer in the 1970s and 1980s. The personal computer is an evolution of the microcomputer on a technical and design level. "By 1977 the pieces were all in place. The Altair's design shortcomings were corrected, if not by MITS then by other companies. Microsoft BASIC allowed programmers to write interesting and, for the first time, serious software for these machines. [...] Finally, by 1977 there was a strong and healthy industry of publications, software companies, and support groups to bring the novice on board. The personal computer had arrived" (Ceruzzi 2003, 240–241). From this moment on, the computer became a complete workstation that became increasingly popular everywhere. The classic example is the Macintosh 128k, launched in 1984.
4. The advent of the Internet in the 1990s. The time for networking had arrived, first with the invention of Ethernet and then with the Internet (Ceruzzi 2003, 291–298). The World Wide Web was born. Important phenomena of these years were the open source movement, of which Linux is an important example, the antitrust trial against Microsoft, and the rise of Google.
5. I want to add a fifth phase, the one we are living in today, characterized by the development of AI (machine learning and deep neural learning) and the availability of large amounts of data. Today we are dealing with complex digital systems that

> are characterized by unpredictability, non-linearity and reversal. The ordering and reordering of systems, structures and networks, as developed in complexity theory, is highly dynamic, processual and unpredictable; the impact of positive and negative feedback loops shift systems away from states of equilibrium. (Elliott 2018, 27)

Now, through these five historical phases, the role of software has become increasingly important, leading to its present dominance. A new type of language has developed and imposed itself, becoming the most widespread form of language on the planet. This language is the "sap" that our laptops, smartphones, tablets, and so forth feed on. Without software all these tools would be unusable. What does this language express? What does it talk about?

Algorithms are not natural things or phenomena, such as atoms and molecules; they do not exist in nature, although nature has inspired some of them (Printz 2018, 265). An algorithm can be defined as "a self-contained, step-by-step set of operations to be performed in the process of attaining a goal" (Primiero 2020, 69). Algorithms have to be formulated as instructions (programming language), then translated into a set of operations performed by a machine (machine code), and finally into electrical charges. In our laptops, smartphones, tablets, cameras, televisions, and so on, data and algorithms are only one thing; data are connected to algorithms, algorithms to data. Data are strings of binary numbers (1s and 0s) stored in memories according to certain classifications, categories, structures. It would be useless if it were not possible to connect them to an algorithm, that is, a set of definite and explicit operations that obtain that data. That string of numbers can be information if and only if a finite set of definite and explicit operations that can be performed by the machine, and whose result is that same string, can be connected to it. Only under this condition, that string, that number, is called "computable," that is, processable by that machine, which is a Turing machine (see Primiero 2020, Chap. 5).

Software is the language that allows humans to "talk" to the machine in order to transmit certain instructions to it. Software allows us to translate our problems into a machine-understandable language. Without it, the computer would remain a mathematical curiosity. "A computer without software is irrelevant, like an automobile without gasoline or a television set without a broadcast signal" (Ensmenger 2010, 5). However, programming has not always been at the core of the computerization of human society. "The activity known as computer programming was not foreseen by the pioneers of computing. During the 1950s they and their customers slowly realized: first, it existed; second, that it was important;

and third, that it was worth the effort to build tools to help it" (Ceruzzi 2003, 108).

Some historical considerations can help in better understanding the software problem. The first electronic computer, the ENIAC (1945), had two essential characteristics: it was electronic and programmable. It could therefore be reprogrammed (by using punch cards) to perform different tasks. Its successor, the EDVAC, was the world's first stored-program computer. The EDVAC did not distinguish between data and instructions, and this allowed for greater flexibility in programming. It marked a turning point: the discovery that the computer could be programmable launched the computer revolution. I still follow Ensmenger (2010, 6): "By allowing the computer to be perpetually reinvented for new purposes, users, and social and economic contexts, software has transformed what was originally intended primarily as a special-purpose technology—essentially a glorified electronic calculator—into a universal machine that encompasses and supersedes all others [...]. It is software that defines our relationship to the computer, software that gives the computer its meaning."

At the beginning of the 1950s, the important textbook by Wikes et al. (1951), in which for the first time the concepts of compiler and subroutine were introduced, came out. The first real automatic program compiler was written by Grace Hopper in 1951 and was the first real answer to a deep crisis in the software industry (overly high costs, lack of professional paths, flawed products). Thanks to the compiler, programming became more understandable. As I said above, the first broadly used programming language was FORTRAN ("Formula Transition"), which was released for IBM 704 in April 1957. COBOL ("Common Business Oriented Language"), the first programming language for commercial and banking applications and machine-independent, was released in 1961 (Ceruzzi 2003, 92–93). "Talking" with machines was getting easier.

Despite this, programming was not considered an important and interesting task in the early computer era. Hardware was privileged over software. This prejudice influenced the development of computer culture in the 1940s and 1950s. Coding—the translation of engineers' work into machine-understandable language—was seen as a static process that could be managed and built by low-level clerical workers; it was considered a mechanical job and was mainly done by women. A certain way of reconstructing the history of the computerization of Western society also contributed to the spread of this idea. In many of these accounts, programming is represented as a subdiscipline of formal logic and mathematics, and its origins are identified in the work of Alan Turing and John von Neumann. However, "this purely intellectual approach to the history of programming conceals the essentially craftlike nature of early programming practice. The first computer programmers were not scientists or mathematicians" (Ensmenger 2010, 32).

With the development of information technology and the transformation of business companies in the late 1960s, the complexity of coding and the need for professionalization of programmers came out. The type and the range of skills required of programmers changed and expanded. "The nascent computing professions were so pressed for resources that they had little time to construct the institutional framework required to produce and regulate software development" (Ensmenger 2010, 19). By the middle of the 1960s, the expenses in information technology were dominated by software maintenance and programmer jobs. Programmer quickly became one of the most requested and highly paid jobs. A seminal date was 1968, when the North Atlantic Treaty Organization (NATO) Conference on Software Engineering took place. This conference set an agenda that has influenced almost all subsequent commercial computing developments. It "started to make the transition from being a craft for a long-haired programming priesthood to becoming a real (software) engineering discipline" (Campbell-Kelly et al. 1996, 201). From that moment, software became capable of guiding the computerization of human society.

Today, the revolution is accomplished. Software shapes the digital universe in which we move and live. Software "invaded" the world, becoming an ecosystem within which we all reside. It is an ecosystem made of billions of written lines. Thanks to this Tower of Babel, everything has become part of Turing's Cathedral (Dyson 2012). These billions of written lines are the result of the systematic work of millions of analysts, engineers, and programmers, and they connect millions of humans, machines, and processes. This infrastructure incorporates and redefines the entire human world from a social, cultural, economic, political, and geopolitical point of view.

However, a contradiction lies here. Although software is the ecosystem in which we are perpetually immersed and that we have created by ourselves, defining software is extremely complicated. There are several different methodological approaches to studying software: information theory, design, computer science, media studies, software engineering, philosophy of technology, sociology of science, systematics, project management, and so on. Nevertheless, the nature of software remains a mystery. "Software is something like a machine, and something like mathematics, and something like language, and something like thought, and art, and information…but software is not in fact any of those other things" (Sterling 1992, 31). "Software is perhaps the ultimate heterogeneous technology. It exists simultaneously as an idea, language, technology, and practice. Although intimately associated with the computer, it also clearly transcends it" (Ensmenger 2010, 8). Hardware is the tangible machine and software is the set of instructions that makes the machine operate in specific ways; "but difficulties quickly set in. Does the distinction apply to computers only or to any machine? Or will we call anything a computer if it seems to take instructions? For example, is knob-turning the software of a clock? Are tracks and their switches the software of trains? Is Bach's written score to the *Art of the Fugue*, perhaps with a human interpreter thrown in, the software of an organ?" (Suber 1988, 89). Hence, we refer to software in many different ways: as a language, a mathematical structure, an artifact, a consumer good, a product to be purchased and installed. Or we usually think of software as a set of services for certain activities, or a way of organizing companies, of business. Many programmers think of themselves as mavericks, artists, or even poets.

Poetry or science? What is programming? What kind of training should a programmer have? As Brooks (1987, 7; see also Brooks 1975) wrote, "programmer, like the poet, works only slightly removed from pure-thought stuff. He builds his castles in the air, from air, creating by exertion of imagination." For Brooks this is not a metaphor; the programmer works mainly with imagination and creativity, not necessarily with mathematical structures, and does not necessarily have mathematical training. This is why programming resists all forms of industrial production. However, Brooks' view has been disputed; large companies cannot do business with their imagination and rely on the individual genius. The debate has marked and marks the entire history of software, especially at the beginning, in the 1950s, when the discipline was still young and had to be defined.

## Software Is Eating the World

In 2011 Marc Andreessen, the founder of Andreessen Horowitz (AH Capital Management), wrote:

> In short, *software is eating the world*. More than 10 years after the peak of the 1990s dot-com bubble, a dozen or so new Internet companies like Facebook and Twitter are sparking controversy in Silicon Valley, due to their rapidly growing private market valuations, and even the occasional successful IPO. With scars from the heyday of Webvan and Pets.com still fresh in the investor psyche, people are asking, "Isn't this just a dangerous new bubble?" I, along with others, have been arguing the other side of the case. (I am co-founder and general partner of venture capital firm Andreessen-Horowitz, which has invested in Facebook, Groupon, Skype, Twitter, Zynga, and Foursquare, among others. I am also personally an investor in LinkedIn.) We believe that many of the prominent new Internet companies are building real, high-growth, high-margin, highly defensible businesses.[3]

*Software is eating the world* in every sense: economically, socially, culturally—everything today is software-mediated. In his article, Andreessen describes this turning point as not only economic, but cultural.

> Today's stock market actually hates technology, as shown by all-time low price/earnings ratios for major public technology companies. Apple, for example, has a P/E ratio of around 15.2—about the same as the broader stock market, despite Apple's immense profitability and dominant market position (Apple in the last couple weeks became the biggest company in America, judged by market capitalization, surpassing Exxon Mobil). And,

> perhaps most telling, you can't have a bubble when people are constantly screaming "Bubble!" But too much of the debate is still around financial valuation, as opposed to the underlying intrinsic value of the best of Silicon Valley's new companies. *My own theory is that we are in the middle of a dramatic and broad technological and economic shift in which software companies are poised to take over large swathes of the economy.* More and more major businesses and industries are being run on software and delivered as online services—from movies to agriculture to national defense. Many of the winners are Silicon Valley-style entrepreneurial technology companies that are invading and overturning established industry structures. Over the next 10 years, I expect many more industries to be disrupted by software, with new world-beating Silicon Valley companies doing the disruption in more cases than not. (emphasis added)

What's happened? What kind of revolution are we experiencing? Is it a real revolution? Andreessen claims that six decades into the computer revolution, four decades since the invention of the microprocessor, and two decades into the rise of the modern Internet, all of the technology required to transform industries through software finally works and can be widely delivered at global scale.

> Over two billion people now use the broadband Internet, up from perhaps 50 million a decade ago, when I was at Netscape, the company I co-founded. In the next 10 years, I expect at least five billion people worldwide to own smartphones, giving every individual with such a phone instant access to the full power of the Internet, every moment of every day. On the back end, software programming tools and Internet-based services make it easy to launch new global software-powered start-ups in many industries—without the need to invest in new infrastructure and train new employees. In 2000, when my partner Ben Horowitz was CEO of the first cloud computing company, LoudCloud, the cost of a customer running a basic Internet application was approximately $150,000 a month. Running that same application today in Amazon's cloud costs about $1500 a month. With lower start-up costs and a vastly expanded market for online services, the result is a global economy that for the first time will be fully digitally wired—the dream of every cyber-visionary of the early 1990s, finally delivered, a full generation later. Perhaps the single most dramatic example of this phenomenon of software eating a traditional business is the suicide of

> Borders and corresponding rise of Amazon. In 2001, Borders agreed to hand over its online business to Amazon under the theory that online book sales were non-strategic and unimportant.

Borders was an international retailer specializing in books and music. In 2011 it declared bankruptcy, appealing to chapter 11. Until a few years ago it was a very important group, but tied to a traditional business model. It has been bypassed by Amazon, which focused its attention solely on the Internet. Everything (health, nutrition, finance, entertainment, transport, construction, socializing, etc.) is turning out to be software-mediated. All services are assimilated by and translated into software. Software has always been present—implicit in every form of digital technology—but is only now taking over.

> Today, the world's largest bookseller, Amazon, is a software company—its core capability is its amazing software engine for selling virtually everything online, no retail stores necessary. On top of that, while Borders was thrashing in the throes of impending bankruptcy, Amazon rearranged its web site to promote its Kindle digital books over physical books for the first time. Now even the books themselves are software. Today's largest video service by number of subscribers is a software company: Netflix. How Netflix eviscerated Blockbuster is an old story, but now other traditional entertainment providers are facing the same threat. Comcast, Time Warner and others are responding by transforming themselves into software companies with efforts such as TV Everywhere, which liberates content from the physical cable and connects it to smartphones and tablets. Today's dominant music companies are software companies, too: Apple's iTunes, Spotify and Pandora. Traditional record labels increasingly exist only to provide those software companies with content. Industry revenue from digital channels totaled $4.6 billion in 2010, growing to 29% of total revenue from 2% in 2004. [...]

Andreessen claims that companies in every industry need to assume that a software revolution is coming, even those industries that are software-based today. Great incumbent software companies like Oracle and Microsoft are increasingly threatened with irrelevance by new software offerings like Salesforce.com and Android. The revolution described by

Andreessen raises many philosophical questions. Software is everywhere. But what is software? Programming languages? Algorithms? "Software is, or should be, a notoriously difficult concept. Historically unforeseen, barely a thing, software's ghostly presence produces and defies apprehension, allowing us to grasp the world through its ungraspable mediation" (Chun 2013, 3). Defining software is not easy. It requires a new perspective on language and the relationship between language and world. There is no single approach to software. It is both a tangible and intangible thing, "a paradoxical combination of visibility and invisibility, of rational causality and profound ignorance" (Chun 2013, 59).

## There Is No Software: Kittler Versus Manovich

Chun's position is symptomatic of a broader issue within media studies. Two of the most important scholars in media studies (Kittler and Manovich) enter a paradoxical debate, in which they assert that there is no software and there is only software, respectively.

In a 1995 paper, Kittler declared: "There is no software." If I opened my laptop, I would find neither programs nor numbers. Software is visible and invisible at the same time. Within the laptop, there are only electrical circuits and voltages: a certain arrangement of electrons. Indeed, what I could see if I looked inside my laptop would not tell me anything about how the machine works. This is why software is an illusion—the illusion of immateriality—and everything is reducible to hardware. According to Kittler, if the computer were not immersed in a human linguistic context dominated by natural language, there would be no need for HLL programming, but everything would happen at the machine level, as a simple flow of electrical impulses. From this point of view, "the signifiers [could] be considered as voltages [...] ultimately everything in a digital computer [can be reduced] to changes in voltages" (Hayles 2005, 45).

Kittler writes:

> Programming languages have eroded the monopoly of ordinary language and grown into a new hierarchy of their own. This postmodern tower of

> Babel reaches from simple operation codes whose linguistic extension is still a hardware configuration passing through an assembler whose extension is that very assembler. As a consequence, far reaching chains of self-similarities in the sense defined by fractal theory organize software as well as hardware of every writing. What remains a problem is only the realization of these layers which, just as modern media technologies in general, have been explicitly contrived in order to evade all perception. We simply do not know what our writing does. (Kittler 1995, 148)

According to Kittler, the software illusion is caused especially by the strange trend in computer culture to obscure the role and weight of hardware. The physical machine is forgotten to make room for programming languages, but without the former, the latter would make no sense. For this reason, Kittler suggests the creation of a "non-programmable system," that is, a form of hardware so advanced that it does not need a language for organizing operations, "a physical device working amidst physical devices and subjected to the same bounded resources." In such a device, "software as an ever-feasible abstraction would not exist anymore" (Kittler 1995, 154).

Kittler's thesis was broadly criticized. In *Software Takes Command*, Manovich claims, "There is only software." According to Manovich, the development of the computer as a meta-medium, that is, a medium capable of connecting, reproducing, and transforming all previous media, is not due to the transition from analog to digital, but to the diffusion of programming languages. Software is more powerful than hardware because of its ubiquity. The same software can run on several different machines, devices, and platforms. "There is no such thing as 'digital media'. There is only software—as applied to media data," says Manovich. But what is software? Software is just "a combination of data structure and set of algorithms" (Manovich 2013, 207–208). "We can compare data structures to nouns and algorithm to verbs. To make an analogy with logic, we can compare [data and algorithms] to subject and predicates" (Manovich 2013, 211; see Manovich 2001, 218–225). However, how can the simple combination of data and algorithm produce so much culture and imagination? How can this combination be "eating the world"?

In my view, many media studies scholars are often victims of their own approach. This approach focuses too much on the role of software in communication and its effects on the sociological and cultural level, without tackling seriously the underlying philosophical questions, such as the following: How does our concept of truth change through software? How does language change through software? Many arguments are too rhetorical and useless. Furthermore, these works are "often written without engagement with the history of technology literature" (Haigh 2019, 15). Although his theses have a remarkable originality, in this paper, Kittler remains too tied to the anti-hermeneutic controversy and his objectives are not clear—his approach is defined as "post-hermeneutics" (Sebastian and Geerke 1990). In Chun and Manovich, I see two inverse processes. Chun, on the one hand, stresses too much the paradoxical aspect of software, often attributing to marginal aspects an excessively problematic nature. Manovich, on the other hand, remains too tied to a definition of software that is flattened on the notion of algorithm, without seeing its intrinsic complexity and multistability. The essence of software is lost in the multiplicity of media. I think that a hermeneutic approach to software, such as the one that will be proposed in this book, can resolve this conflict of interpretations and propose a model of understanding the dynamic nature of software.

Of course, I am not saying Kittler's work is useless. Kittler is one of the most interesting philosophers of technology. He claims, "There is no software," but at the same time he is a programmer and an important software theorist (Marino 2020, 161–197). In his book *Gramophone, Film, Typewriter* (1999), he is very close to one of the theses I defend in this book, namely the connection between software and the history of writing.

Kittler is a media theorist deeply influenced by French poststructuralism, particularly by Foucault. The transition from orality to writing, and from the Gutenberg Galaxy to modern sound and image recording techniques, to digital technologies, is a crucial theme in *Discourse Networks* (Kittler 1985). *Gramophone, Film, Typewriter* develops this project by examining how information and communication change in the post-Gutenberg era, in the era of electronic media. Kittler argues that today the predominance of the alphabetical order and symbolic mediation has been replaced by the predominance of the materiality of the sign, that is, by non-symbolic recording methods such as cinema and the gramophone. For Kittler, the Lacanian triad (symbolic, imaginary, and real) reflects this media revolution: the typewriter belongs to the world of the symbolic, the film to that of the imaginary, while the gramophone to that of the real (Kittler 1999, 15). In the passage from one world to another there is an ever wider disconnection between meaning and reality. In this passage, the human being is gradually overcome: Kittler calls upon the anti-humanist rhetoric of Nietzsche, Foucault, and Lacan. The human being as a cognitive subject is surpassed by microprocessors, as is software by hardware. Honestly I disagree with this trend, which often seems like just a way to avoid facing problems and accept easy causal chains—for example, in Kittler there is also an interpretation of the history of technology unilaterally based on war and military industry. One of the main theses of this book is that even the materiality of writing is hermeneutic, that is, it creates meaning and interpretation—this is what I try to demonstrate through the historical analysis of the symbolic revolution of the seventeenth century in mathematics (see Chap. 4). I argue that a regression from the meaning to the materiality of the sign, of the trace and of the electric charge connected to it, takes place in software. However, this regression does not eliminate the meaning as such. Another essential point that distinguishes me from Kittler, and media studies in general, is that, in my opinion, software is not just a matter of communication. Digital technology is a complex ecosystem in which human beings live and which redefines all previous activities and relationships.

## Software Circumscribes the Planet

There is another aspect that makes it very hard to formulate a good definition of software. Nowadays, software is redefining not only the economic and commercial world, but also architecture, design, and the entire global geopolitical structure. Software has become a planetary phenomenon, a mega-infrastructure that conveys new forms of mobility, sovereignty, power, and control. In this global mega-infrastructure, most interactions are machine/machine and not human/machines. In other words, through software, computation has ceased to be a simple mathematical structure and has become a new global ecosystem run by powerful learning systems and deep neural networks, whose behavior is often difficult to understand and predict even for the engineers who designed them (Voosen 2017). This means that today we can no longer think of software as a simple program written and tested by a team of programmers and run on a machine. Software came out of the box, and now it redefines the box and all its surrounding environment. It is an overwhelming process. It cannot be described just as the building of a new environment, that is, a new *Umwelt* that humans have made for themselves. The philosophical analysis of software cannot underestimate this crucial aspect.

In *The Stack*, Benjamin Bratton described this planetary software infrastructure through the concept of "stack," that is, as a set of six different levels or platforms: Earth, Cloud, City, Address, Interface, and User. Bratton offers us "a portrait of the system we have but perhaps do not recognize, and an antecedent of a future territory," "a larval geopolitical architecture" that may serve as a "metaplatform of an alternative counter-industrialization" (Bratton 2016, 5–6). Based mainly on Deleuze (1997), Galloway (2004), and Virilio (1991), Bratton's model intends to include all technological systems as part of a singular planetary-scale computer, updated to reflect and interpret the demands of the Anthropocene era. "Bratton may thus be seen as an advocate for the idea that media theory is the successor discipline to metaphysics. […] His ambitions, however, extend much further still, as the basis for a normative project for how media theory might also be of practical use in addressing an issue no less pressing than global anthropogenic climate change" (Tuters 2017).

Bratton turns to Carl Schmitt's argument that political epochs are founded on new spatial divisions (Schmitt 2003). Building on a tradition of thought extending through Machiavelli, Locke, Vico, Rousseau, and Kant, Schmitt saw land-appropriation as the ontological root from which all subsequent judgments derive in normative orders of history and politics. For Schmitt (2005), sovereignty signifies the autonomy of "the political" since it was the sovereign who decided on that which was undecidable by law—this is the "state of exception." Schmitt recognized, however, that technology diminished territorially derived sovereign power by decoupling political jurisdiction from geographic knowledge. A practical example of this is how the establishment of a global common datum, through technologies that form the basis of contemporary geolocation services, came to replace national cadastral surveys, thereby stabilizing borders while at the same time making them easier to ignore from the perspective of those in command of the technology. Bratton refers to this phenomenon as "superjurisdiction" (Bratton 2016, 120), offering the example of how Google increasingly operates with the force of a global sovereign, as, for instance, when it becomes involved in inter-border conflicts—between Nicaragua and Costa Rica, between Tibet and Kashmir, and between Israel and Palestine. He proposes the 2009 Google-China conflict as potentially signifying an epistemological rupture with an idea of sovereignty derived from territorial occupation and categorical juridical identification toward a new notion of "platform sovereignty" (Bratton 2016, 22). Bratton speculates on how the apparatus of planetary-scale computing might operate in terms of a "full-spectrum governmentality" (Bratton 2016, 101), automating and radically erasing the usual modern sovereign system, from the scale of the molecular to that of the atmospheric. The Stack is "a machine literally circumscribing the planet, which not only pierces and distorts Westphalian models of state territory but also produces new spaces in its own image" (Bratton 2016, 52).

How does The Stack work? What is it made from? Each platform of The Stack is an extremely complex organization that includes machines, programs, human agents, communication networks, and implies new forms of sovereignty, institutions, and rules that cannot be reduced to those of the modern state system. "The Stack is an accidental megastructure, one that we are building both deliberately and unwittingly and is in turn building us in its own image. […] It is not 'the state as a machine' (Weber) or the 'state machine' (Althusser) or really even (only) the technologies of governance (Foucault) as much as it is *the machine as the state*. Its agglomeration of computing machines into platform systems not only reflects, manages, and enforces forms of sovereignty; it also generates them in the first place" (Bratton 2016, 5–8). The platforms act in a centralized and decentralized, local and global, abstract and concrete way, and so define a new geopolitics. Software is the fuel that drives this immense technological and political network: "[Stack] relies on software as both a kind of language and a kind of technology, of algorithms of expression and the expression of algorithms, and its twisting of the conceptual and the machinic can sometimes bring emotional distress" (Bratton 2016, 57). The Stack is the Moloch of our time.

The levels in The Stack interact in a vertical manner. Each *User* (human and non-human) has a specific position in The Stack, and from this position it activates an *Interface* in order to manipulate things, actions, and interactions. "Human and nonhuman *Users* are positioned by The Stack (perhaps rudely) as comparable and even interchangeable through a wide-ranging and omnivorous quantification of their behaviors and effects. The preponderance of data generated by users and the traces of their worldly transactions initially over trace the outline of a given *User* (e.g., the hyperindividualism of the quantified self-movement), but as new data streams overlap over it and through it, the coherent position of the *User* dissolves through its overdetermination by external relations and networks" (Bratton 2016, 71). The *User* is first "a grotesquely individuated self-image, a profile," but as the same process is "oversubscribed by data that trace all the things that affect the *User*, now included in the profile, the persona that first promises coherency and closure brings an explosion and liquefaction of self" (Bratton 2016, 71).

All the actions of the *User* are translated into the *Interface*. "Contemporary human being is primarily a being in interaction, which constantly manages digital interfaces" (Vial 2013, 208). *Interface* means space. In the *Interface*, each element has a position, an address, which allows them to interact with other structures or agents on the same level or on another. This *User-Interface* connection is the condition of possibility of the datafication of experience. "The *Address* layer examines massively granular universal addressing systems such as IPv6 (Internet Protocol version six) (including cryptographically generated hash addresses), which would allow for a truly abyssal volume of individual addresses. Such individuated addresses make any thing or event appear to the *Cloud* as communicable entity, and for The Stack, computation then becomes a potential property of addressed objects, places, and events, and a medium through which any of these can directly interact with any other" (Bratton 2016, 70). The *Cloud* ensures storage and communication; it is a vast archipelago of servers and communication infrastructures: server farm, massive database, optical cables, wireless transmission media, distributed applications, and so forth.

The *City-Earth* pair corresponds to the *User-Interface-Cloud* triad, which represents the physical side of The Stack. The layer *City* "comprises the environment of discontinuous megacities and meganetworks that situate human settlement and mobility in the combination of physical and virtual envelopes" (Bratton 2016, 70). The *Earth* is the physical layer that supports the entire architecture of The Stack, providing energy at every level. "There is no planetary-scale computation without a planet, and no computational infrastructure without the transformation of matter into energy and energy into information. [...] *The Stack terraforms the host planet by drinking and vomiting its elemental juices and spitting up mobile phones*" (Bratton 2016, 75–83). There is an ancestral materiality of software and computation: they come from the earth in order to return to the earth—almost like a divinity who returns to earth to "save" humans and deliver them a new world!

I chose to carefully analyze Bratton's works because with them emerges a broader philosophical and political problem posed by software. Software is not a neutral object but includes in itself a series of *normativities*, namely, a set of rules that have to be understood and interpreted. This is an important aspect of the hermeneutical problem we mentioned above. As planetary infrastructure, software interprets us before we interpret it. The condition of the possibility of the interpretation of software is the interpretation of the users that the software gives. It is a double interpretation: we can analyze software only through the software itself, but software involves a series of normativities that pre-condition and pre-interpret us. Interpreting software is therefore first and foremost a gesture of reflection that must look at software normativities, that is, at how the software interprets us.

In his analysis of the technological system, Winner (1993) invites us to see artifacts as materialized laws that redefine how we can act for generations to come. Lee and Larsen (2019) formulate the concept of *algorithmic normativities*: "[...] Algorithms are intertwined with different normativities, and these normativities come to shape our world. [...] A crucial task for the researcher thus becomes to analyze invisible algorithmic normativities to understand how they are imbued with power and politics" (Lee and Larsen 2019, 2–3). Numerous scholars have stressed the importance of the negotiations between programmers during the construction of algorithms (Ziewiitz 2017); others have highlighted the relationship of mutual adaptation between algorithms and human beings (Larsen and Dudhwala 2019); "humans adapt to algorithmic outputs, an adaptation that includes normative ideas about how the outcomes of algorithms are interpreted as normal or abnormal" (Lee and Larsen 2019). Larsen and Dudhwala (2019) claim that the output of algorithms trigger, or as they propose, *recalibrate* human responses. Grosman and Reigeluth (2019) distinguish three types of algorithmic normativities: (a) technical, (b) socio-technical, and (c) behavioral. But we can also include a fourth level: that of imagination.

Simondon (Beaubois 2015) has developed a theory of technical imagination that profoundly renews the Kantian concept of schematism. "Imagination is not only the faculty of inventing or giving birth to representations outside of sensation; it is also the ability to perceive in the objects some qualities that are not practical, that are neither directly sensorial nor entirely geometric, which do not relate either to pure matter or to pure form, but which are on this intermediate level of the schemes" (Simondon 2012, 73–74). Imagination, for Simondon, is not simply a faculty internal to the subject, a mental capacity, but a property of things, that is, their structural and functional analogies. These analogical processes are called "schemes" (a term that Simondon takes from Berson). The technical object arises from the ability to grasp and develop the "operational schemes" already present in things. The engineer therefore (1) understands the operational schemes and (2) elaborates a new operational scheme, the project, which defines the principles of machine operation. The project is essentially a diagram that can be translated into different contexts and applied to different objects. The project "stands" in the machine and represents its profound nature. Furthermore, the transformation of the diagram can lead to the discovery of new schemes and new operating methods. Thanks to its theoretical and practical nature, the diagram is a heuristic tool.

The project is only a certain type of imagination, the technical imagination. Digital technologies can also create new forms of imagination; they are "imaginative machines" (Romele 2019; Finn 2017). The book *The Imaginary App* assembles contributions from scholars, artists, and independent researchers to analyze apps through a diversity of approaches, including media analytic, sociological, philosophical, and psychoanalytic. The collection is thematically organized into four sections: Architectures, Prosthetics, Economics, and Remediations. Each section contains several relatively short chapters and as a fun contrast to the theoretical rigor established within the primary texts, 15 color plates reproduced from the volume's accompanying art exhibition populate a gallery of literal imaginary apps. These apps range from the fantastical, such as Christoffer Laursen Hald's teleportation app "Telephort," to the conceivable, such as Alexander Chaparro's Wikilution, "a network for the sharing and dissemination of information that can aid protests" (Miller and Matviyenko 2014, 273). These artistic interpretations of what imagined future users may expect from apps draw attention to some of the tacit expectations inherent to software. Software is the product of a design process and "design does not want to produce materials, but create better experience regimes" (Vial 2013, 203).

The hermeneutical perspective presented in this book helps, in my opinion, to shed light on all these transformations (imaginative, political, and social) that software conveys. "Code increasingly shapes, transforms, and limits our lives, our relationships, our art, our culture, and our civic institutions. It is time to take code out from behind quotation marks, to move beyond execution to comment, to document, and to interpret" (Marino 2020, 54).

## Notes

1. "Some phenomena of the world are not properly speaking *phenomena*, in the sense that they do not appear, do not manifest themselves to us in the perceptive experience. Located in a perceptive afterworld, they are invisible and almost make us believe that they do not exist. These phenomena are for example […] quantum phenomena, or rather, we should say *quan-*

*tum noumena*. Kant defines the noumenon as something that is outside the field of the sensible experience. And this is what happens with the quantum world, this world 'hidden' for physicists, which is 'essentially mathematical', and which, for this very reason, can never be seen" (Vial 2013, 143–144, my translation).

2. I take up this concept of philosophical problems and philosophical work because I consider it compelling, though not exhaustive. I share the view according to which "philosophy is not conceptual aspirin, a super-science, or the manicure of language either. Its method is conceptual design, that is, the art of identifying and clarifying open questions and of devising, refining, proposing, and evaluating explanatory answers. [...] Philosophy is, after all, the last stage of reflection, where the semanticization of Being is pursued and kept open" (Floridi 2019, 26). However, I do not share the framework of Floridi's investigation, namely, a philosophy of information.
3. See: https://a16z.com/2011/08/20/why-software-is-eating-the-world/.

## References

Bachelard, G. 1933. *Les intuitions atomistiques*. Paris: Boivin.

———. 1934. *Le nouvel esprit scientifique*. Paris: Puf.

———. 1938. *La formation de l'esprit scientifique*. Paris: Puf.

Beaubois, V. 2015. Un schématisme pratique de l'imagination. *Appareil*. journals.openedition.org/appareil/2247.

Beaune, J.-C. 1980. *L'automate et ses mobiles*. Paris: Flammarion.

Bratton, B. 2016. *The Stack: On Software and Sovereignty*. Cambridge, MA: MIT Press.

Brooks, F. 1975. *The Mythical Man-Month: Essays on Software Engineering*. New York: Addison Wesley.

———. 1987. No Silver Bullet: Essence and Accidents of Software Engineering. *IEEE Computer* 20 (4): 10–19.

Campbell-Kelly, M., Aspray, W. 1996. *Computer: A History of the Information Machine*. New York: Basic Books.

Ceruzzi, P. 2003. *History of Modern Computing*. Cambridge: MIT Press.

Chun, Wendy Hui Kyong. 2013. *Programmed Visions. Software and Memory*. Cambridge, MA: MIT Press.

Deleuze, G. 1997. Postscript on Control Societies. In *October: The Second Decade, 1986–1996*, ed. R. Krauss. Cambridge, MA: MIT Press.

Dyson, G. 2012. *Turing's Cathedral. The Origins of the Digital Universe*. London: Penguin Books.
Elliott, A. 2018. *The Culture of AI*. London; New York: Routledge.
Ensmenger, N. 2010. *The Computer Boys Take Over*. Cambridge, MA: MIT Press.
Finn, E. 2017. *What Algorithms Wants. Imagination in the Age of Computing*. Cambridge, MA: MIT Press.
Floridi, L. 2019. *The Logic of Information*. Oxford University Press.
Frabetti, F. 2015. *Software Theory*. London; New York: Rowman & Littlefield (Media Philosophy).
Galloway, A. 2004. *Protocol. How Control Exists After Decentralization*. Cambridge, MA: MIT Press.
Gille, B. 1978. *Histoire des techniques*. Paris: Gallimard.
Goody, J. 1977. *The Domestication of the Savage Mind*. Cambridge University Press.
Grosman, J., and T. Reigeluth. 2019. Perspectives on Algorithmic Normativities: Engineers, Objects. *Activities. Big Data & Society* 6 (2): 1–6.
Günther, G. 1957. *Das Bewusstsein der Maschinen: eine Metaphysik der Kybernetik*. Krefeld: Agis-Verlag.
———. 1962. *Cybernetic Ontology and Transjunctional Operations*. See: http://vordenker.de/ggphilosophy/gg_cyb_ontology.pdf.
Haigh, T., ed. 2019. *Exploring the Early Digital*. Berlin: Springer.
Havelock, Eric. 1963. *Preface to Plato*. Cambridge: Cambridge University Press.
Hayles, C. 2005. *My Mother Was a Computer. Digital Subjects and Literary Texts*. Chicago: University of Chicago Press.
Huyghe, P.-D. 2002. *Du commun. Philosophie pour la peinture et le cinema*. Belval: Circé.
Ihde, D. 1991. *Technology and the Lifeworld*. Indiana University Press.
Khanna, Parag. 2016. *Connectography. Mapping the Future of Global Civilisation*. New York: Random House.
Kittler, F. 1985. *Discourse Networks 1800/1900*. Stanford University Press.
———. 1995. There Is No Software. http://raley.english.ucsb.edu/wp-content/Engl800/Kittler-nosoftware.pdf
———. 1999. *Gramophone, Film, Typewriter*. Stanford University Press.
Larsen, L. Björklund, and F. Dudhwala. 2019. Recalibration in Counting and Accounting Practices: Dealing with Algorithmic Output in Public and Private. Big Data & Society (1)1–12.
Latour, B. 2005. *Reassembling the Social. An Introduction to Actor-Network Theory*. Oxford University Press.

Lee, F., and L. Björklund Larsen. 2019. How Should We Theorize Algorithms? Five Ideal Types in Analyzing Algorithmic Normativities. Big Data & Society. (1) 1–6.

Leroi-Gourhan, A. 1964. *Le geste et la parole 1. Technique et langage*. Paris: Albin Michel.

Manovich, Lev. 2001. *The Language of New Media*. Cambridge, MA: MIT Press.

———. 2013. *Software Takes Command*. London: Bloomsbury.

Marino, M. 2020. *Critical Code Studies*. Cambridge, MA: MIT Press.

Miller, P., and S. Matviyenko. 2014. *The Imaginary App*. Cambridge, MA: MIT Press.

Ong, W. 1982. *Orality and Literacy*. London; New York: Methuen.

Primiero, G. 2020. *On the Foundations of Computing*. Oxford: Oxford University Press.

Printz, J. 2018. *Survivrons-nous à la technologie?* Paris: Les acteurs du savoir.

Romele, A. 2019. *Digital Hermeneutics*. London; New York: Routledge.

Sack, W. 2019. *The Software Arts*. Cambridge, MA: MIT Press.

Salanskis, M. 2011. *Le monde du computationnel*. Paris: Encre marine.

Schmitt, C. [1950] 2003. *The Nomos of the Earth in the International Law of the Jus Publicum Europaeum*. New York: Telos Press Publishing.

———. [1922] 2005. *Political Theology: Four Chapters on the Concept of Sovereignty*. Chicago: University of Chicago Press.

Sebastian, T., and G. Geerke. 1990. http://soundartarchive.net/articles/Sebastian-1990-Kittler-Discourse_Networks.pdf.

Serfati, M. 2005. *La revolution symbolique*. Paris: Ed. Petra.

Simondon, G. [1958] 2005. *L'individuation à la lumière des notions de forme et d'information*. Paris: Millon.

———. 2012. *Du mode d'existence des objets techniques*. Paris: Aubier.

Sterling, B. 1992. *The Hacker Crackdown*. New York: Bantam Books.

Suber, P. 1988. What Is Software?. https://legacy.earlham.edu/~peters/writing/software.htm.

Tuters, M. 2017. Scenario Theory—Review of Benjamin H. Bratton. *The Stack: on Software and Sovereignty. Computational Culture* 6 (28th November 2017). http://computationalculture.net/scenario-theory-review-of-benjamin-h-bratton-the-stack-on-software-and-sovereignty/.

Vial, S. 2013. *L'être et l'ecran*. Paris: Puf.

Virilio, P. 1991. *The Lost Dimension*. New York: Semiotext(e).

Voosen, Paul. 2017. The AI Detectives. As Neural Nets Push into Sciences, Researchers Probe Back. *Science* 357: 22–27.

Wikes, M., D. Wheeler, and S. Gill. 1951. *Preparation of Programs for an Electronic Digital Computer*. Reading, MA: Addison-Wesley.
Winner, L. 1993. Upon Opening the Black Box and Finding It Empty: Social Constructivism and the Philosophy of Technology. *Science, Technology, & Human Values* 18 (3): 362–378.
Ziewiitz, M. 2017. A Not Quite Random Walk: Experimenting with the Ethnomethods of the Algorithm. *Big Data & Society*.: 1–13.

# 2

# Software and Lived Experience

## Introduction

Computers and software are some of the most important parts of our everyday life. Our IT appliances are indispensable assets for most contemporary individuals. We use IT appliances for communicating, watching movies, paying our taxes and bills, and shopping. "This new revolution is arguably more profound than the previous ones, and we are just beginning to register its initial effects" (Manovich 2001, 19). We now share with ICT (Information & Communication Technology) equipment a new kind of space: the informational environment or the *infosphere*. "The global space of information, which includes the cyberspace as well as classical mass media such as libraries and archives…brings upfront the need to reinterpret Man's position in reality—that is Man's position in the infosphere" (Russo 2012, 66).

This work demonstrates that software-based experience, namely the experience mediated by digital technology and therefore by software, is based on a radical break between the code and "lived experience" or the

This chapter is a new development of Possati (2020).



L. M. Possati, *Software as Hermeneutics*,
https://doi.org/10.1007/978-3-030-63610-4_2

experience of ordinary users throughout the use of ICT appliances.[1] This break has deep consequences for our way of *perceiving* the world. As Luciano Floridi (2010; see also 2011) claims, the digital revolution brings up questions about the relations between *physis* and *techne*, respectively understood on the one hand as nature and reality, and on the other hand as technique and technical practices. "Information technologies are creating altogether new *e*-nvironments that pose new challenges for the understanding of us in the world" (Russo 2012, 66). While Floridi claims that the digital revolution affects the position and the role of human beings as ethical agents, I want to emphasize that this revolution affects first of all the position and the role of human beings as perceiving and epistemic agents. If the experience becomes digital, its unity becomes problematic. I think that in order to understand the origins of the aforementioned break (with no negative connotation), we must ask ourselves another more thoughtful question: what is software?

The chapter is organized as follows. I start (section "Assassin's Creed Odyssey, a Case Study") by analyzing a case study: the video game *Assassin's Creed Odyssey*. A video game is a relationship between a human being (the user) and a machine that can redefine all aspects of the world's experience pertaining to the user. Analyzing the video game experience is useful because video games are one of the most extreme and profound—and common—*digital* experiences. In section "Code Versus Lived Experience," I examine the gap between code and lived experience. I claim that the code's concealment is a necessary condition of the digital experience. *The code must be absent*: when I either play a video game or look at a picture on my laptop screen, I do not see the code, I do not even know what code is creating that experience. In order to know the code and understand it, I need a specific knowledge and a technique. However, *the code is the source of my digital experience*. Reaching the core of the break between code and lived experience, in section "Software, This Unknown Land," I discuss the ontological definition of software as an entity. I argue that the origin of the phenomenological changes introduced by software in the world lies in the ontology of software. Software—I claim—is a complex object, composed of many different technical, semiotic, and cultural levels, whose unity is problematic. In the last part of the chapter, section "Design as Imaginary Act: An Attempt of

Solution," I argue that the break between lived experience and code is recomposed by imagination through the act of design.

## *Assassin's Creed Odyssey*, a Case Study[2]

The video game *Assassin's Creed Odyssey* (Ubisoft 2018) follows protagonist Layla Hassan, who also led the previous game in the long-running series: *Assassin's Creed Origins*. The game opens with the Battle of Thermopylae. The Spartan king Leonidas leads his soldiers against the Persians of Xerxes. After this initial scene, the game jumps to the "present" day. During an archeological expedition, Layla, an agent of the Assassin Brotherhood, finds the ancient spear that belonged to Leonidas and, with the help of a friend, Victoria Bibeau, manages to extract the DNA of a brother and a sister who lived in the fifth century BC during the Peloponnesian War, Alexios and Kassandra. Layla must choose which brother's story to follow through the game's mechanic of the Animus, which allows the protagonist to experience past events. Her ultimate goal is to find the stick of Hermes Trismegistus and restore the final order of things.

*Assassin's Creed: Odyssey* is an open-world game in which the most important aspect is not a fixed and static goal (conquer a city, free a princess, kill the enemy, complete a path, etc.), but the exploration of a world—a historical world in this case: that of ancient Greece and the Persian wars. The general narrative plot is articulated in a multiplicity of possible alternative stories, love relationships, travels, encounters with other characters, and knowledge of new parts of the game world. Ancient Greece is represented in a mythical, rarefied way. The game offers a vast territory that can be explored freely. The player is placed into a setting that they can know in great detail, potentially encompassing hundreds of hours of gameplay. They can also interact with their surroundings in many ways and can engage in combat with any person or animal, using a vast arsenal of weapons. The player can also entertain other types of relationships with the characters, changing the game's plot. Different stories are intertwined. There is neither a fixed script nor predetermined order. The player—who guides the character from a "third person" perspective, where the "camera" is positioned above and behind the character being controlled[3]—writes the story by choosing the actions to be performed. It is a shared experience: the machine offers a context full of possibilities that it is up to the player to filter and interpret.

The most interesting characteristic of the game is that there are many development possibilities. There are two kinds of interactions within the game offered to the player: (a) movements, that is the possibility of moving in the *Odyssey* world: crossing valleys, islands, caves, beaches, lakes, half-destroyed temples, cities, forests, and so on; (b) actions, namely the possibility of carrying out one action instead of another: killing (above all), loving, making friends, having sex, stealing money, or conversing with philosophers, politicians, or priests. The story is amplified also with new parallel narratives provided by the developers through the game's website, such as the stories "The Lost Tales of Greece: Daughters of Laila" and "A Poet's Legacy." It is not a situation comparable to a book or a film that presents closed, defined narrative evolutions. *Odyssey* is an open-world game, which offers a radically different experience compared to other cultural objects. Other examples of open-world games include *Subnautica* (Unknown World Entertainment 2018), *Kingdom Come: Deliverance* (Warhorse Studios 2018), *The Witcher* (CD Projekt RED 2007), *Mass Effect* (Electronic Arts 2007), *Just Cause* (Avalanche Studios 2006), and *Mad Max* (Avalanche Studios 2015). Rules and constraints are present in open-world games—for instance, goals, challenges, and intermediate tests—but their influence is much lower than in other games. As a player, I can also choose not to fight, save my enemies and go for a walk in an unknown forest or swim in the sea. In this case, the main narrative plot does not go on, but I cannot know if in that forest or in that stretch of the open sea there are no other secondary stories or deviations from the main story, or simply completely unexpected visual experiences.

The open-world video game is not simply a story, an object, a film, a toy, or a program, but an all-encompassing experience replacing—or trying to replace—reality and directly involving the user or users. In this case, digital technology creates new environments, and "those environments are at once either cognitive—in the sense of the space of knowledge—and applied—in the sense of the space of *application* of such a knowledge" (Russo 2012, 67). This totalizing experience, in some cases, even comes to challenge the user as a subject. For example, while multiboxing (using more than one computer to simultaneously control multiple characters within the same game world), a single player can play multiple roles, impersonate more characters, and engage the game world from multiple perspectives.

This all-encompassing experience poses a philosophical issue. As Mathieu Triclot (2013, 18) writes: "where are we when we play a video game? [...] we are in an intermediate zone" (translation mine). Triclot refers to Winnicott and his concepts of "transitional objects" (Winnicott 2005). The experience of the game—the play—lies neither in the subject nor in the machine, but in their interaction. Defining this intermediate area is very hard because it requires the analysis of many different aspects: the pleasure produced by the play, the consumer dimension, the user's involvement in terms of senses, body, and emotions, and the mutual adaptation of software and user.

Playing is above all fun. As Roger Caillois (1958) explains, following Johan Huizinga (1951),[4] the experience of fun is based on four essential and irreducible impulses mixed up in different proportions: competition, chance, simulation, and vertigo (fear of defeat, bankruptcy, death, etc.). These four categories are not fixed labels, but they form a polarized, moving space. A game is a free activity, separated from the rest, which develops within a specific space, where the player's relationship to the four categories are dynamic and ever-changing. The video game summarizes, takes to the extreme, and transforms Caillois' categories thanks to digital technology. Triclot (2013) defines the relationship between player and machine in the video game a *hallu-simulation*: the player finds themself in an artificial world where every aspect is the result of a code, of software, except for the player themself. A *hallu-simulation* is a paradoxical combination between the freedom of fun and the determinism of an algorithm.

> The video game produces a strange and unknown form of calculation: *a calculation that produces a dream, a dream woven by a calculation.* [...] *The video game is a little dream dust through which capitalism wakes up from its*

> *great dream; things that are dreams grafted onto computational machines.* (Triclot 2013, 42, emphasis added)

Video games allow an experience that is detached from our daily experiences. However, this experience has important heuristic effects on our existences. In video games (at least in most of them), the suspension of the relation to the "actual" world aims to transform our existence. Through this suspension, *hallu-simulations* can also convey ideological content. For this reason, a political critique of video games is necessary.

The alternative universe produced by software is not purely mechanical or deterministic. The machine does not eliminate all chance, but rather controls it, and this affects the player's emotional reaction. In *Tetris* (Pazitnov 1984), for example, the number of types of shapes or blocks is more limited than the number of ways these shapes or blocks can appear. This difference produces the concealment of the finiteness of the combinations, making the blocks appear on the screen. The player is led to believe that these combinations are infinite, uncontrollable, and unpredictable, even if they are not. This illusion is a *hallu-simulation*. An open-world game uses a similar dissimulation effect to produce in the player a feeling of freedom and improvisation. This is even more evident in some masterpieces like *Blueberry Garden* (Svedäng 2009) or *Minecraft* (Persson 2009) which present a living and dreamlike world, constantly evolving, apparently without a purpose, but very involving (Triclot 2013, 40–41). The former, designed by Erik Svedäng, who uses the music of composer Daduk, comes with the appearance of an old silent, sweet, and melancholic film. His visual style recalls Klee's paintings. You cannot die, there are no enemies. The purpose of the game is to take care of a huge garden: to grow fruits and plants, and protect and move them. You can walk or fly from one place to another: the setting is infinite. The player has to repair a leaking water mixer which threatens to submerge and destroy the garden. Despite—or perhaps thanks to—very simple 2D graphics, *Blueberry Garden* manages to transform the fun of the game into a different experience, becoming pure art, pure creativity. This reveals a different problem: understanding what kind of image modification the digital technology causes, as well as its role in art, cinema, and visual culture (Mitchell 1986).

## Code Versus Lived Experience

*Assassin's Creed Odyssey* is a useful tool to define the central issue that I am investigating. A video game is a relationship between a subject and a meta-medium, the computer (Manovich 2013), that can go so far as to redefine all aspects of the individual's world perceptions. An open-world game defines another world. My question is the following: how is this experience possible? Is there a *digital* experience? In short, I want to understand what is an "experience after software," or a "software-sized experience."[5]

The "softwarization" of the world is one of the most global and complex phenomena of the last thirty years (1990–2020). If we want to understand contemporary society, our analysis cannot be complete until we consider software as a main cultural (and philosophical) factor. Today every part of our life, not only communication, has been affected by software. *Software never sleeps*: Google, Facebook, Amazon, Twitter, eBay, and the other main IT brands change their algorithms daily, often via automatic processes. Some software change themselves by themselves in a limited way: they have an autonomous "life" materialized in many servers all over the world. This "life" is the core of the global economy, culture, social life, and politics, but "this 'cultural software'—cultural in a sense that it is directly used by hundreds of millions of people and that it carries 'atoms' of culture—is *only the visible part of a much larger software universe*" (Manovich 2013, 7, emphasis added). Therefore "software is the invisible glue that ties it all together [...]. *software as a theoretical category* is still invisible to most academics, artists, and cultural professionals interested in IT and its cultural and social effects" (Manovich 2013, 8–9). It is a new kind of lived experience requiring an entirely new philosophical approach.

The relationship between games and "traditional" media is essential, here. "[T]he computer media revolution affects all stages of communication, including acquisition, manipulation, storage, and distribution; it also affects all types of media—texts, still images, moving images, sound, and spatial constructions" (Manovich 2001, 19). Software is the core of this revolution: "[it] shapes our understanding of what media is in general" (Manovich 2013, 121). The digitalization of media does not have the effect of negating the differences between various kinds of media (photography, literature, art, cinema, theaters, etc.). Software, instead, creates new forms of organization between existing media and thus (1) transforms the characteristics of these media and (2) creates new media. The computer is therefore a *metamedium*, "simultaneously a set of different media and a system for generating new media tools and new types of media. In other words, a computer can be used to create new tools for working with the media types it already provides as well as to develop new not-yet-invented media" (Manovich 2013, 102–3). The way in which Photoshop or Instagram revolutionized photography is indisputable (Manovich 2016a, b). In a more radical way, we can say that "*all media techniques and tools available in software applications are 'new media'—regardless of whether a particular technique or program refers to previous media, physical phenomena or a common task that existed before it was turned into software*" (Manovich 2013, 141, emphasis added). The phenomenon recently affecting the new media the most—says Lev Manovich—is hybridification, or the fusion of languages that on one hand threatens the autonomy of individual media, and on the other hand gives rise simultaneously to new media (Manovich 2013, 167–81). An example of hybridification is the language of visual design (graphic design, web design, interface design, motion graphics, etc.).

By transforming our media, software modifies our experience of the world. In this redefinition, however, software opens up a radical and profound fracture. The thesis of this chapter is that this fracture challenges the unity of experience and the status of digital experience. To solve this issue, we must ask ourselves another question: *what is software?*

Let us go back to our case study. *Assassin's Creed Odyssey* produces a very particular simulation of a physical experience. The play presents two layers: first is the code, an autonomous type of writing, ceaselessly reading and modifying itself; and second is the lived experience, that is, the interaction between the user and the *metamedium* depending on the effects of the code (in both physical and abstract sense). I say "the interaction between the user and the *metamedium*" and not "the interaction between the user and the machine" because the *metamedium* cannot be understood as a physical machine. A *metamedium* may involve many physical machines, but it is not defined by those devices. The physical devices matter the least to the game experience; I can have nearly the same experience (play my video game) on different devices.

Therefore, there is a difference between the code—that is, the algorithm, which is a string of characters, a writing[6]—and the relation between the game and the users—that is, the play itself. The first level is invisible: the computational process is hidden.

One can consider the asymmetry between the discrete level of screen pixels and the visual experience one has when they play their video game, which is continuous and holistic (Salanskis 2011, 113–14). This is an effect of digitalization consisting of two steps: sampling and quantization. First, an image is translated into a set of pixels: it is fragmented into discrete pieces. "Sampling turns continuous data into discrete data, that is, data occurring in distinct units: people, the pages of a book, pixels" (Manovich 2001, 28). Secondly, pixels are quantified; "[each pixel] is assigned a numerical value drawn from a defined range (such as 0–255 in the case of an o bit greyscale image)" (Manovich 2001, 28). Sampling and quantization are the two main stages of distance created by the code.

Let us make this more explicit. I see this picture. I have a visual experience (Fig. 2.1).

This is the algorithm *behind* the picture (Fig. 2.2):

When I see this picture, I do not see the code but the result of the algorithmic performance interacting with me. When I read a text on Microsoft Word, I do not read the code, but I have *a hallu-simulation*. I live a new kind of experience. Like in a video game, I am in a fictional context (the screen page). "[I]n the case of 'electronic paper' such as a Word document or a PDF file, we can do many things which were not possible with ordinary paper: zoom in and out of the document, search for a particular phrase, change fonts and line spacing, etc. [...]" (Manovich 2013, 287). This performance is hidden; the computational genesis is concealed. The occultation of the code-performance is a necessary condition of the digital experience. Software is a noumenon, "a phenomenon without phenomenality" (Vial 2013).

> In this perspective, a noumenon is a phenomenon without phenomenality: [it is a phenomenon] that does not phenomenalize, that does not manifest

**Fig. 2.1** Image from David Burnett, *Schizo 2 in JS*

> itself, that does not appear to us, that does not reach the 'lived world', that we do not perceive—a kind of unmanifested manifestation. The quantum phenomena of nuclear physics are such noumena. (Vial 2013, 53—translation is mine)

The code is hidden not simply because it is invisible. Writing and reading software are fundamentally incommensurable to the experience of looking at an image. When I play a video game, I live a single and consistent experience of fun and involvement. And yet, "behind the screen" there is something completely different: a string of ones and zeroes. Where does the synthesis come from? Learning to play does not mean learning the algorithm. I can even become a champion without knowing the code at all, nor the possible different algorithms that "stand behind" the screen. I could also observe on another screen, simultaneously, the code that produces the game as I play. However, I would not get much. If I were not a programmer, that string would not tell me anything—it could be a Balzac novel, I could not distinguish the two. Furthermore, even if I were a programmer, I could not be sure that it is precisely that string of code that runs the game because the same game scene—the sketch121526

```
int penx = 0;
int peny = 0;
int de = 2;
cLista lista = new cLista();
int el0 = 0;

int by;

// ------------------------------------------

void setup() {
  size(800, 600);
//  size(1024, 768);
//  size(screen.width, screen.height);
  smooth();
  frameRate(30);
  colorMode(HSB, 1);
  ellipseMode(CENTER);
  stroke(0, 0, 0, 0.7);
  background(0, 0, 1);
//  strokeWeight(0.3 * de + 1);
strokeWeight(2);

  city();
}

// ------------------------------------------

void draw() {
  int t = 0;
  while((t++ < 50) && (el0 < lista.nr)) {
    lista._draw(el0++);
  }
```

**Fig. 2.2** The code of the image. Find the complete code on: https://openprocessing.org/

same effect—can be caused by very different algorithms, and the same software can also refer to different algorithms. Even if I were the creator of the game, I could not be sure that the string I see on the screen next to the one which I am playing on is the string that produces that game. It could be an old version of the program that I created, which has meanwhile evolved thanks to other programmers.

The image of the buildings I showed earlier is neither a photograph nor a painting: there is no relationship with reality, neither before nor after the production of the image. The image of the buildings is totally created by the software running on my desktop. While photography (analog) and the picture exceed the world of signs (code, language), the digital image is the product of signs—it is only code. An analog photograph is not codified, but rather adheres to human vision. For example, both in the eye and in photography there is a relationship with natural light. A digital photograph need not adhere to human vision but rather need only simulate this adhesion; it is simulated vision.[7] In the digital case, "we end up having a file containing an array of pixels holding color values, and a header file specifying image, dimensions, color profile, information about the camera and shot conditions such as exposure, and other metadata." In other words, we end up with strings of numbers, but "*unless you are a programmer, you never directly deal with these numbers*" (Manovich 2013, 151, emphasis added). The code is invisible; it is understandable only through a specific technique—as if it was a sort of unconscious, more indecipherable than the human one.[8] We need to use specific knowledge and technique to translate the strings of numbers into contents acceptable to our senses and understanding.

We can now better formulate the problem: what holds together these two opposite sides—the algorithmic manipulation and the lived experience—of just one single event? Why and how does the user interpret those electrical stimuli produced by software in that precise way? How can software, despite this break, produce a coherent experience and have a cultural role? The essentially Kantian question of the unity of our experience is at stake.

Someone may object though: the break between software and lived experience does not tell us anything new. The same difference between mechanism and lived experience is also found in many other technical objects. All technology is based on what we might be called an illusion of transparency and immediacy (Van Den Eede 2011). There are many different theories of technology transparency and opacity, including Bruno Latour (2005a, b) whose Actor-Network Theory is a radical redefinition of science and technology itself. For Latour, facts and technologies are black boxes: once their complex design and production process is completed, this background disappears. Technologies are at the same time *opaque*—because we have forgotten their functioning and cultural background—and *transparent*—because we can freely use them. A very similar idea can also be found in Don Ihde's post-phenomenology (1990). For example, when I drive a car, I experience the journey, but I do not see the engine. I can know nothing about mechanics and still live that experience. When I turn on a light in my apartment, I do not have to be an electrician or know the house's electrical system to live the experience of seeing the light. The same concept can be applied to a computer. The break between software and lived experience takes a typical feature of many technical tools to its extreme.

I argue that this objection is actually based on a misunderstanding of the nature of software. Software is different from other technologies. In software, if we open the black box, we find nothing in it.

## Software, This Unknown Land

In responding to the objection, I want to highlight some crucial aspects:

(a) Software is not a tool. It is not a mechanism such as a car engine or an electrical system. Behind software, *there is culture*: a style of writing, an aesthetic vision, a social perception, a commercial objective. Software is a cultural system (cf. Geertz 2013), that is a network of meanings. This is the thesis of Ed Finn (2017), who establishes a new "algorithmic reading" by creating concepts such as "algorithmic imagination" or "algorithmic aesthetics." For Finn the algorithm is not simply manipulation of symbols separated from their meanings; it is not a purely syntactic concept. On the contrary, the algorithm has not only consolidated cultural and historical roots, but also significant consequences on the way in which human beings act and conceive of themselves and their world. The algorithm "is an idea that puts structures of symbolic logic into motion. It is rooted in computer science, but it serves as a prism for a much broader array of cultural, philosophical, mathematical, and imaginative grammars" (Finn 2017, 41). That is why the algorithm is a "culture machine… [which] operates both within and beyond the reflexive barrier of effective computability, producing culture at a macro-social level at the same time as it produces cultural objects, processes, and experiences" (Finn 2017, 34). According to Finn, we must distinguish four different levels of interpretation of the concept of algorithm: (a) the computational reading, the pure mathematical structure theorized by Alan Turing, Alonzo Church, and Kurt Gödel, and which can be summarized by the concept of *effective computability*; (b) the cybernetic reading, developed by Norbert Wiener, in which the algorithm becomes the instrument for the control of information and communication applied to society, biology, and physics; (c) the symbolic reading, according to which the algorithm uses language to produce effects in the world; (d) cognitive reading, where language is the first instrument of the externalization of a thought, which modifies that very thought, also known as the "embodied mind thesis" (Clark 2014), or the relationship between cognition and its tools.

(b) Software is a living organism, and its life goes beyond the machine and the individual user experience. Software is an autonomous environment with its own rules. Software often updates itself and works and develops without human intervention. Engineers often cannot explain the behavior of their own systems. "Wall Street traders give their financial 'algos' names like Ambush and Raider, yet they often have no idea how their money-making black boxes work" (Finn 2017, 16). Web pages are automatically generated from databases by using templates created by web designers; furthermore, "information about the user can be used by a computer program to customize automatically the media composition as well as to create elements themselves" (Manovich 2001, 37).

(c) Software has an artistic dimension. It is not just a combination of logical-mathematical signs and operations. The beauty of software is a crucial element: the more beautiful software is, the more it works and is alive. Software designers are artists and strongly claim this role (see Sack 2019). The pioneers of digital media such as Sutherland, Nelson, Kay, Agroponte, and Engelbart have underlined the importance of the artistic factor in their inspiration (see Wardrip-Fruin and Montfort 2003). Of course, "culture" is behind all technological artefacts. Technologies are the result of cultural transmission; they have specific styles and aesthetic visions. I claim here that software has always a tendency to go beyond its technical and logical-mathematical nature, and that this tendency produces a particular form of cultural content.

Given these three aspects, how does the opacity of software differ from that of other technologies? Let us imagine a human being traveling in their car. At some point the car breaks down. They stop and open the car's hood: what do they see? The engine. With technical knowledge, they could understand the car's functioning and, perhaps, how to repair it. In any case, the driver has, or can have, a direct perceptual experience of machine operations: combustion, petrol, internal mechanisms, and so on. They can disassemble the machine piece by piece and *show* the direct connection between movement and engine.

Now let us consider another human being while using their laptop. Let us suppose that they decide to open the laptop and look inside: what do they see? They do not see the machine working, but rather a cluster of circuits that are only partially informative about how the laptop works. This is the point: the actual functioning escapes continuously. Let us suppose that they are a programmer and know perfectly the software that runs on their computer. However, the connection between code strings and machine operations varies continuously. Let us suppose—recalling our previous example—that they could simultaneously monitor the electric impulses in the computer, the data stored in the memory, and the code strings. Even in this case, they could not directly experience *the connection* between strings, data, and electrical impulses. *The user must presuppose this connection but cannot perceive and/or show it.* Even if they dismantle the laptop, they will not find this connection. *The user cannot show why an abstract symbol produces a physical effect.*

The reason for the discrepancy between code and lived experience is not in our ignorance. It is structural; it is a material *a priori*. Software is not the computer's instruction booklet, a technical knowledge detached from the machine. Software *is* the operation of the computer (or at least, an essential part of it). Software writing *takes place* in the computer. It is *performative*: it does what it "says" (writes/re-writes). Thus the illusion of transparency and immediacy of technology turns into a paradox.

So, then, *what is software?*[9] We must understand the specific nature of software to understand the break between code and lived experience.

Niklaus Wirth defined software/programs as "algorithms plus data structure" (Wirth 1976, 1). This definition is extended by Manovich, who argues that the intellectual work of programming consists "of two interconnected parts: [first,] creating the data structures which logically fit with the task which needs to be done and are computationally efficient and [second,] defining the algorithms which operate on these data structures." Thus, software is "a combination of a data structure and set of algorithms" (Manovich 2013, 207–8). To make a comparison with language, "we can compare data structures to nouns and algorithm to verbs. To make an analogy with logic, we can compare them to subject and predicates" (Manovich 2013, 211, see also 2001, 218–25).

Is this first definition satisfactory enough for us? Since it is a descriptive definition, not an ontological or metaphysical one, we must go further.

From an ontological point of view, software is an enigma. Timothy R. Colburn (1999) defines software as a "concrete abstraction," underlining software's dual—almost contradictory—nature. Abstractness and concreteness are essential properties of software. Colburn develops this thesis through two arguments:

(a) Computer programming is not a branch of pure mathematics—mathematics deals with formal abstractions, while programming deals with abstractions that do not eliminate the content but widen it, making it more expressive.
(b) There is a deep distance between the binary language and the physical states of the corresponding machine.

> [T]he characterization of physical state in machine language as zeros and ones is *itself* an abstraction; the kinds of physical states that this language 'abstracts' are boundless. They may be the electronic states of semiconductors, or the state of polarization of optical media, or the states of punched paper tape, or the position of gears in Babbage's 18th century analytical engine. (Colburn 1999, 16)

This process of abstraction widens even more as Colburn describes how "the whole history of computer science has seen the careful construction of layer upon layer of distancing abstractions upon the basic foundation of zeros and ones" (Colburn 1999, 16).

Colburn argues that we cannot reduce either argument to the other: a monistic approach is useless. On the other hand, a dualistic approach cannot be causal: how can a string of numbers and operations give rise to a physical state? Colburn prefers to speak of "pre-established harmony" (Colburn 1999, 17), namely a parallelism between code and machine established not by God, but by the programmer. "Programmers today can live almost exclusively in the abstract realm of their software descriptions, but since their creations have parallel lives as bouncing electrons, theirs is a world of concrete abstractions" (Colburn 1999, 18).

Colburn's conclusion is criticized by Nurbay Irmak (2012) who defines software as an "abstract artifact:" an abstract object, built by the human mind for a precise purpose. Software has no spatial characteristics, but it is placed in a time, as well as in a historical period—it is created and can be destroyed, unlike Platonic ideas. Irmak compares software with music: "I think that both software and musical works are abstract artifacts which are created by software engineers or composers with certain intentions," and therefore "computer programs are not types and thus the relation between computer programs and their physical copies cannot be understood in terms of the type/token distinction" (Irmak 2012, 70).

Nevertheless, Irmak's conclusion also seems problematic or incomplete. In which sense can the following statement "musical works and software come into existence by some human being's act of creation" (Irmak 2012, 71) help us understand the ontological status of software? Irmak's thesis, although true, does not help us answer this question.

The whole problem must be reconsidered. Computer programming is a long and complex process articulated throughout numerous phases. There are two key concepts to be considered: problem solving and coding. The first is the more creative part of the job. As many programmers say, "the sooner you start coding your programs, the longer it is going to take," thus "first think, code later." Thus, understanding and defining the problem to be solved is the core of programming. We must know exactly what we want to do, outline the solution—organizing tasks, subtasks, units, and so forth—and then select and design the algorithm. Even the procedure of writing an algorithm correctly is not easy—even something as fundamental as specifying how a value will be assigned to an operation. The coding-programming depends on problem solving. The choice of programming language will be dictated by the nature of the problem (building an Internet site, making an app, etc.) and the languages available on the computer in use.

Therefore, I call software a specific entity defined by (a) a problem, (b) an algorithm (logic + control),[10] and (c) a code (a formal language). By this I do not mean that the only way to define software is a functional or instrumental one. This is an elementary conceptual "cell" through which to build our definition of software. I argue instead that a *basic* form of software ontology—software as *entity* before other technical, artistic, cultural, and social determinations—should start from these three categories as connected parts of a unique entity. In other words, I claim that these three categories are three variables whose unity is complex.

Let us try to clarify the matter by using this diagram, inspired by Victor Papenek (1975) (Fig. 2.3):

First, none of these levels, taken separately, is software. Second, these levels make software happen provided they work together as a whole. Software is the dynamic synthesis of all three variables. I want to stress the presence of Yin-Yang symbols, through which the profound complementarity between and within all the phases of the process is emphasized.

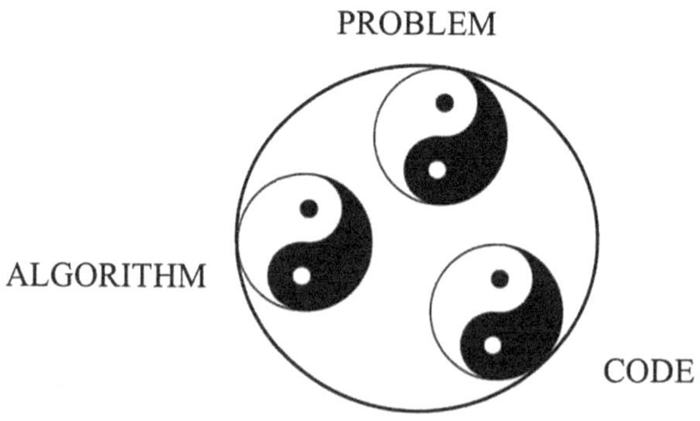


**Fig. 2.3** The three fundamental aspects of software

Software is not the algorithm. The same algorithm can be written in many different types of languages: there is no direct relationship. "If an algorithm is a set of steps we can carry out ourselves, programming is the activity by which we write down the steps in the notation that the computer understands. Then it is the computer that will carry them out" (Louridas 2020, 23). Obviously, the algorithm is essential because it is *what is written*—it is the meaning, the proposed solution. However, software cannot be identified with a single programming language. It is not a single high-level or a low-level language: we can in fact have different high-level or low-level languages connected to the same software, or vice versa. Likewise, we cannot state that software is just the machine code, the pure binary code, even if it represents the heart of the machine. By itself, a string of ones and zeroes has no meaning. Essentially, "a programming language gives us a way to describe to a computer the steps of algorithms. It also provides the means to structure them using the three fundamental control structures: sequence, selection, and iteration" (Louridas 2020, 23).

To be even more precise, we should distinguish the algorithm from the computational process. Not every computational process implements an algorithm. The program that runs for ever online, for example, is not an algorithm but a computational process. An algorithm must have a finite number of steps that must be precise (we can perform them without confusion) and must produce a result. An algorithm does something; it exists because it does something useful. Moreover, "the algorithm must be effective. A human should be able to execute each step in a reasonable amount of time with pen and paper" (Louridas 2020, 26). This is a very important point: the concept of algorithm belongs to practical philosophy, not to mathematics. "Most early computer scientists were mathematicians, and computer science uses a lot of mathematics, but it is not a mathematical discipline" (Louridas 2020, 27).

Another clarification is needed: the notion of computation does not exactly correspond to that of program. In the 1930s, Church and Turing argued that there is a family of programs capable of spelling out any rule for manipulating digits. Compared to the general set of computations, however, only a small subset contains the computations that can be translated into a program (a set of instructions that can be executed by a computer). "Computers can only perform computations for which there are programs. By definition, the only computations for which there are programs are defined over a domain that has at most countably many entities—for instance, strings of letters from a finite alphabet. In addition, the vast majority of functions from strings of digits to strings of digits are such that there is no program for computing them" (Piccinini and Scarantino 2016, 24).

To clarify this point, I propose to distinguish three levels:

1. The pure level of calculus, which is physical. This is what is computing in terms of a Turing Machine. There is no semantics at all, even a formal one: logical or mathematical semantics is already an abstracted view of what is going on in a Turing Machine. The bits, that is the 0s and 1s are only tokens considered for their distinguishability and their position on the Turing memory band. The fact they are named 0 and 1 is not essential at this level.
2. The formal level of the mathematical interpretation of the calculus level. This is the abstracted view adopted by computer scientists and logicians to understand the behavior of machines, their limitations, and so forth. This level can also be called functional.
3. The interpretative level of the algorithm, where semantics is borrowed to a non-formal domain, comes from the application for which the software is designed.

At the calculus levels, code has no meaning; it consists of inputs that trigger computations (movement on the memory band) as physical causes. Code, at the machine level, has no semantics or meaning; this level cannot be fully accounted for by mathematical definition (e.g., lambda calculus). The programmer gives code a meaning through the algorithm.

The diagram I proposed earlier overlooks many technical details; for example, there may be intermediate levels of language. However, the basics of the chapter contents have been portrayed in there.

In such a diagram, each step is defined as an interpretation: the algorithm is translated into the high-level programming language, which in turn must be translated into the machine language through low-level languages. The machine language, in turn, must be translated into a series of physical states: that is the execution. The problem is the relationship between these levels: Does the translation have a causal scope? Can an algorithm—an abstract entity—give rise to a physical state? If we accept Colburn's position, the programmer would be a sort of *deus ex machina* who optimizes the level's harmony upon request. Yet often the programmer cannot know what really happens, cannot follow the life of software, and is unaware of its developments.

Software is a complex artifact, consisting of a set of elements, irreducible to any one of them. Its unity cannot be reduced to a sum of the parts.

This definition is not entirely satisfying. Coding is the dominant element in the diagram. But what kind of coding are we talking about? What is the relation between coding and writing? What is the principle of the unity of software levels? Defining software is a very hard problem. I wanted here only to draw a basic definition, emphasizing how software is a layered object, that is, composed of many different levels of analysis, some symbolic, others physical. There is not a single software ontology, but many software ontologies; we need a plural, critical ontology. Plurality and asymmetry explain the paradoxical nature of software.

This ontological enquiry improves our phenomenological approach toward how we experience digital technologies. The fracture between software and lived experience cannot be assimilated to the usual technological opacity, and this by virtue of the nature of software. There is no software on my laptop, as Friedrich Kittler says (1995). Nevertheless, software—like the Kantian noumenon—is the condition of possibility of the experience that I live when I use digital devices. Thus, what is the principle of synthesis between software and lived experience? If the operation of the computer depends on the synthesis of software levels, *what makes the synthesis happen?* The break between code and lived experience depends on the symbolic/physical fracture in software. They are intertwined questions.

## Is Software Ubiquitous? The Borderline Between Hardware and Software

Is the hardware the mediation we are looking for? Is the machine physicality a bridge between the user-subject and software? I intentionally leave the question partially open. I think that the softwarization of our society is changing the relation between hardware and software. Software is ubiquitous. The same software can run simultaneously on multiple machines. The experience produced by the machine—playing a video game, writing a text, and more—can be caused by different types of software, with different styles and structures. Today hardware is modeled by software. The most powerful algorithms in the world—Google PageRank, for example—affect not only our research but also the Internet evolution, the construction of the devices to access it, the cognitive practices related to its contents, and all this in a potentially unlimited way—see the concept of "permanent extendibility" of the computer as metamedium (Manovich 2013, 337).

However, software needs hardware. Software evolution has not cancelled hardware relevance. A digit is a paradoxical entity: it is physical status with a symbolic function. How can these two features be together? What is physical computation? According to Russ Abbott (2019), neither Church nor Turing considered this question. Abstract structures are thoughts in the mind of some human being. But they are ineffable, that is, they cannot be displayed or exhibited. We can use a language to express them, but there will always be a distance:

> [...] a lambda calculus expression is not the same as the process of applying that lambda calculus expression to an argument and evaluating the result. This is similar to saying that the execution of a computer program is not the same as either (a) the program itself or (b) a characterization of how the execution of the program works [...] the process of thinking cannot (yet?) be captured and offered up as fixed 'thing' to be examined. (Abbott 2019, 7)

Analyzing this question would be fascinating. Yet this is not what I intend to do here.

## Design as Imaginary Act: An Attempt of Solution

Let us go back to our main issue: the break between software and user experience. I want to propose a solution. I believe that the mediation between software and lived experience must be sought in a more subtle and complex concept, one lying in *design*. In the act of design, imagination[11] responds to the break produced by software. In all digital experience's levels, from the construction of software to the graphic interface (GUI) and to the physical machine, design produces the unity of digital experience—and the unity of software as well.

> The ultimate job of design is to transform man's environment and tools and, by extension, man himself. Man has always changed himself and its surroundings, but recently science, technology, and mass production have advanced so radically that changes are more rapid, more thorough, and often less predictable. (Papenek 1975, 29)

Design is a collective endeavor. In today's experience of software engineering, there is not one designer, but a multitude of designers who often work in collaboration (and sometimes ignore each other). The imaginative process is always multi-layered. Many of these layers involve non-human agents. Moreover, users also take part in this process.

From a philosophical point of view, design is a hermeneutical act which interprets and changes our experience. Stéphane Vial (2013) argues that design transforms our experience in three ways: *ontophanique*, *callimorphique*, and *socioplastique*. The "design effect" is not a logical consequence of a syllogism, but rather a phenomenological effect (Vial 2017). First, a design object transforms our sensorial perception. Vial calls this effect *effet ontophanique*. "The history of technical revolutions is the history of *ontophanique* revolutions… [because] there is no manifestation of the world outside the technical conditions in which phenomena are possible" (Vial 2013, 21, translation mine). A stone, for example, will appear to me differently if I interact with it in a pre-mechanical technical system or in a digital system, by using a laptop. The perceived object does not actually change: the stone remains always the same. The act of perceiving is transformed. There is therefore a direct relationship between the technical system and the perceptive mode. In that respect, for instance, the telephone's main innovation could be analyzed as the separation of the voice from the physical presence, to be heard without being seen. This is an *effet ontophanique*. We could even say that "every experience of the world depends on a technical *ontophanie*" (Vial 2013, 20). Just as Bachelard spoke of *phénoménotechnique*, inviting us to think of the technical equipment used by the scientist as the prerequisite for the existence of the scientific phenomenon, Vial claims that design is "a general structure of perception conditioning a priori the way in which beings appear" (Vial 2013, 20). Perception is always built, amplified, and transformed by the overall system of devices which we cope with (telephone, camera, typewriter, laptop, etc.).

Vial also mentions another design effect: the *effet callimorphique*, or how design produces the perception of formal beauty. Design means creating harmonious forms, with a character, an expression, an empathy, and therefore responding to or evoking a primary human need.

Lastly, there is the *effet socioplastique*, or how design produces and shapes social interactions. Design always encapsulates a vision about a new possible world. The forms created by a designer, unlike those of art, have a public use, a public life. More generally, design is always driven by the ambition for a new society, a transformation of the society.

Design is a phenomenological revolution: a revolution of our way of perceiving reality and society. "A phenomenological revolution is produced when the act of perceiving is affected or modified by an artistic, scientific or technical innovation" (Vial 2013, 17). Digital technologies involve an act of design, and thus

> perceiving *in the digital age* means to renegotiate the act of perception, in the sense that digital entities force us to forge new perceptions [...]. There is nothing natural about this perceptual renegociation [*rénégociation perceptive*]. It demands from the contemporary subject a real phenomenological work to learn to perceive this new category of entities, the digital beings, whose phenomenality is unprecedented. (Vial 2013, 17)

The phenomenological revolution in the digital age is a psychic and social event; it consists in "*learning to perceive digital beings for what they are*, without metaphysical superstructure [...]" (Vial 2013, 17, emphasis added).

To be clear, design is present at every phase of the digital experience. There are four fundamental levels: (a) software design, (b) graphics design, (c) web design, and (d) hardware design. The purpose of software design is to adapt the code as much as possible to the problem to be solved and the user's needs. Graphics design combines symbols, images, and texts in order to create the Graphical User Interface (GUI), which has to align as closely as possible to the user's expected perceptive experience. As Wendy Hui Kyong Chun writes,

> interfaces offer us an imaginary relationship to our hardware: they do not represent transistors but rather desktops and recycling bins. Interfaces and operating systems produce 'users'—one and all. Without OS there would be no access to hardware; without OS there would be not actions, no practices, and thus no user. Each OS, in its extramedial advertisements, interpellates a 'user': it calls it a name, offering it a name or image with which to identify. So Mac users 'think different' and identify with Martin Luther King and Albert Einstein; Linux users are open-source power-geeks, drawn to the image of a fat, sated penguin (the Linux mascot); and Windows users are mainstream, functionalist types perhaps comforted [...] by their regularly crashing computers. (Chun 2013, 66–67)

The field of computer graphics is immense and involves different sectors like data visualization, 3D modeling, dialogue design, image manipulation and storage, animation, lighting sources, and shading techniques (see Foley et al. 1997).

Web design focuses more on web navigation and therefore must also take into account other parameters such as site visibility, advertising, and marketing. It is linked to the vast field of SEO (search engine optimization).

The fourth level is the hardware design. Let us think about the revolution introduced, for example, by the introduction of joysticks in video games. "The Atari VCS was the first cartridge-based system to come with a joystick controller. Although joysticks were already in use in arcades by 1977, the introduction of the VCS joystick into the context of the home undoubtedly did much to popularize the controller" (Montfort and Bogost 2009, 22). The joystick fundamentally shaped the experience of gaming with the Atari VCS, producing an experience unlike other modes of gaming, such as text interfaces. This revolution has ramifications even in modern gaming as joysticks (or the modern equivalent) are included in the controllers for all modern consoles and are a common option for non-console computer games as well.

Together these design levels interpret and modify our perceptive, aesthetic, and social experience, to answer the challenge of the break between software and lived experience. Thanks to these design levels, that users do not experience code phenomenologically becomes irrelevant—namely, users do not feel the need to look for software, although software is in fact the condition for their experience. The answer to the break between software and lived experience is thus an act of design, that is, an act of imagination.

## Notes

1. By "lived experience" I refer to Heidegger's concept of "being-in-the-world," "familiarity," that is, "a fundamental experience of the world [...] we do not normally experience ourselves as subjects standing over against an object, but rather as at home in a world we already understand" (Blattner 2007, 43). My approach is phenomenological. I start from the observation of how our common experience—above all our way of perceiving—is transformed by digital technology. In this sense, I use the expressions "lived experience" and "users experience" as synonyms insofar as digital technologies have become part of our "familiarity."
2. For the philosophy of video games, see Triclot (2013); Donovan (2010); Juul (2005). For the psychology of games, see Turkle (2005).
3. This point is worth being stressed. The video game produces a powerful form of immersion in the image, very different from that of cinema or theater. For example, in a video game there is no editing. The cinematographic image exceeds the code, while the image of the video game is a

product of the code. On the problem of identification linked to the gaze or the player's vision in the video game, see Triclot (2013, 50ss).

4. I refer to these authors, in particular Johan Huizinga and Donald W. Winnicott, because they are generally quoted in the literature on the game and entertainment experience.
5. I consider these two expressions as synonyms, even if there is a slight difference between them. By "experience after software" I mean the common experience after the advent of softwarization in our society. By "software-sized experience" I mean the experience realized through software, that is, the experience "extended" by software.
6. Here I use the expressions "code," "writing," and "computation" almost as synonyms, although obviously there is a difference. See Catherine Hayles (2005).
7. This is just an example. I do not want to treat here the complex question of the realism in the transition from analog photography to digital photography.
8. I thank the anonymous reviewer for raising an interesting question: why would the computational unconscious be more complex than the psychological one? *Can we talk about an algorithmic unconscious?*
9. My approach is very close to Frederica Frabetti's (2015). I want to problematize software as software, rather than looking at the social and cultural meanings of software. I want to consider software as a specific and autonomous source of meanings; "[…] we do need to shift the focus of analysis from the practices and discourses concerning them to a thorough investigation of how new technologies work, and, in particular, of how software works and of what it does" (Frabetti 2015, xviii). Frabetti thinks (deconstructively) software as an aporetic concept precisely because the point of separation—or of transition—between hardware and software, physical and symbolic, is undeterminable. I completely agree with this point of view: this is my starting point. To my mind, this book is very close to this approach. I claim that software produces paradoxes, it is a paradox. As I will explain, the ontological definition I propose is the first step toward a critical software ontology.
10. For a basic definition of algorithm, see Robert Kowalski (1979). I also want to mention Matthew Fuller (2008, 15): "[An algorithm is] a description of the method by which a task is to be accomplished; the algorithm is thus the fundamental entity with which computer scientists operate. *It is independent of programming languages and independent of the*

*machines that execute the programs composed from these algorithms*. An algorithm is an abstraction, having an autonomous existence independent of what computer scientists like to refer to as 'implementation details,' that is, its embodiment in a particular programming language for a particular machine architecture (which particularities are thus considered irrelevant)" (emphasis added).

11. For the relation between imagination and digital media, see Romele (2018) and Stiegler (2001).

## References

Abbott, R. 2019. The Bit (and Three Other Abstractions) Define the Borderline Between Hardware and Software. *Minds and Machine* 29 (2): 229–285. https://link.springer.com/article/10.1007/s11023-018-9486-1.

Avalanche Studios. 2006. *Just Cause*. Avalanche Studios. Playstation.

———. 2015. Mad Max. Avalanche Studios. Playstation.

Blattner, W. 2007. *Heidegger's 'Being and Time.' A Reader's Guide*. New York, NY: Continuum.

Caillois, R. 1958. *Les jeux et les hommes: le masque et le vertige*. Paris: Gallimard.

CD Projekt RED. 2007. *The Witcher*. CD Projekt RED. Playstation.

Chun, W. 2013. *Programmed Visions. Software and Memory*. Cambridge, MA: MIT Press.

Clark, A. 2014. *Mindware. An Introduction to the Philosophy of Cognitive Science*. Oxford: Oxford University Press.

Colburn, T.R. 1999. Software, Abstraction, and Ontology. *The Monist* 82: 3–19.

Donovan, T. 2010. *Replay. The History of Video Games*. Lewes: Yellow Ant.

Electronic Arts. 2007. *Mass Effect*. Electronic Arts. Xbox One.

Finn, E. 2017. *What Algorithms Want. Imagination in the Age of Computing*. Cambridge, MA: MIT Press.

Floridi, L. 2010. *Information: A Very Short Introduction*. Oxford: Oxford University Press.

———. 2011. *The Philosophy of Information*. Oxford: Oxford University Press.

Foley, J., V. Andries, and S.K. Feiner. 1997. *Introduction to Computer Graphics*. Boston, MA: Addison-Wesley.

Frabetti, F. 2015. *Software Theory*. London; New York, NY: Media Philosophy.

Fuller, M. 2008. *Software Studies. A Lexicon*. Cambridge, MA: MIT Press.

Geertz, C. (1973) 2013. *The Interpretation of Culture*. New York, NY: Basic Books.

Hayles, C. 2005. *My Mother Was a Computer. Digital Subjects and Literary Texts.* Chicago, IL: University of Chicago Press.

Huizinga, J. (1938) 1951. *Homo ludens. Essai sur la function sociale du jeu.* Paris: Gallimard.

Ihde, D. 1990. *Technology and The Lifeworld.* Bloomington, IN: Indiana University Press.

Irmak, N. 2012. Software Is an Abstract Artifact. *Grazer Philosophische Studien* 86: 55–72.

Juul, J. 2005. *Half-Real. Video Games Between Real Rules and Fictional Worlds.* Cambridge, MA: MIT Press.

Kittler, F. 1995. There Is No Software. ctheory.net, October. https://journals.uvic.ca/index.php/ctheory/article/view/14655/5522.

Kowalski, R. 1979. Algorithm = Logic + Control. *Communications of the ACM* 22: 424–436.

Latour, B. 2005a. *La science en action. Introduction à la sociologie des sciences.* Paris: La Découverte.

———. 2005b. *Reassembling the Social. An Introduction to Actor-Network Theory.* Oxford: Oxford University Press.

Louridas, P. 2020. *Algorithms.* Cambridge, MA: MIT Press.

Manovich, L. 2001. *The Language of New Media.* Cambridge, MA: MIT Press.

———. 2013. *Software Takes Command.* London: Bloomsbury.

———. 2016a. Subjects and Styles in Instagram Photography. http://manovich.net/index.php/projects/subjects-and-styles-in-instagram-photography-part-1.

———. 2016b. Subjects and Styles in Instagram Photography—2. http://manovich.net/index.php/projects/subjects-and-styles-in-instagram-photography-part-2.

Mitchell, W. 1986. *Iconology.* Chicago, IL: University of Chicago Press.

Montfort, N., and I. Bogost. 2009. *Racing the Beam. The Atari Video Computer System.* Cambridge, MA: MIT Press.

Papenek, V. 1975. *Design for the Real World. Human Ecology and Social Change.* London: Thames & Hudson.

Pazitnov, A. 1984. *Tetris.* Atari: Atari Games.

Persson, M. 2009. *Minecraft.* Playstation: Mojang.

Piccinini, G., and A. Scarantino. 2016. Computation and Information. In *The Routledge Handbook of Philosophy of Information*, ed. Luciano Floridi, 23–29. London; New York, NY: Routledge.

Possati, L. M. 2020. Software and Experience. A Phenomenological Analysis of Digital Technology. *Hermès* (CNRS).
Romele, A. 2018. Imaginative Machines. *Techné: Research in Philosophy and Technology* 22 (1): 98–125.
Russo, F. 2012. The *Homo Poieticus* and the Bridge Between *Physis* and *Techne*. In *Luciano Floridi's Philosophy of Technology. Critical Reflections*, ed. Hilmi Demir, 65–81. London; New York, NY: Springer.
Sack, W. 2019. *The Software Arts*. Cambridge, MA: MIT Press.
Salanskis, M. 2011. *Le monde du computationnel*. Paris: Les Belles Lettres.
Stiegler, B. 2001. *Technics and Time 3: The Cinematic Time and the Question of Malaise*. Stanford: Stanford University Press.
Svedäng, E. 2009. *Blueberry Garden*. Microsoft Windows/Linux/Mac: Svedäng.
Triclot, M. 2013. *Philosophie des jeux video*. Paris: La Découverte.
Turkle, S. 2005. *The Second Self. Computers and the Human Spirit*. Cambridge, MA: MIT Press.
Unknown World Entertainment. 2018. *Subnautica*. Unknown World Entertainment. Playstation.
Van Den Eede, Y. 2011. In Between Us: On the Transparency and Opacity of Technological Mediation. *Foundations of Sciences* 16: 139–159.
Vial, S. 2013. *L'être et l'écran*. Paris: Puf.
———. 2017. *Le design*. Paris: Puf.
Wardrip-Fruin, N., and N. Montfort. 2003. *The New Media Reader*. Cambridge, MA: MIT Press.
Warhorse Studios. 2018. *Kingdom Come: Deliverance*. Warhorse Studios. Playstation.
Winnicott, D. W. (1971) 2005. *Playing and Reality*. London: Routledge.
Wirth, N. 1976. *Algorithms + Data Structures = Programs*. Upper Saddle River, NJ: Prentice Hall.

# 3

# Can We Interpret the Code? A Ricoeurian Perspective on Software

## Introduction

At the end of the previous chapter, we formulated a hypothesis: design "replies" to the fracture between software and experience by providing mediation between these two poles. Design, we claimed, is a hermeneutic act, and therefore it entails an interpretative process. We have distinguished four levels of this process: (a) software design, (b) graphics design, (c) web design, and (d) hardware design. In this chapter, I intend to develop this thesis. I argue that the four levels of design are rooted in a deeper "hermeneutic ground." The software itself contains a hermeneutic; it is in itself a hermeneutic gesture. Through the analogy between software and text, in Ricoeur's sense, I enrich the initial definition of software formulated in the first chapter.

The aim of this chapter is to contribute to understanding software from a "continental" perspective. The so-called continental philosophy has something to say about software. I argue that Paul Ricoeur's philosophical hermeneutics can help in developing a new understanding of the nature of software, thus fueling the philosophical debate on digital technology. Software can be thought of as text, in the Ricoeurian sense of the term. Software is text; indeed, it is text *par excellence*. This means that



L. M. Possati, *Software as Hermeneutics*,
https://doi.org/10.1007/978-3-030-63610-4_3

software takes to the extreme a conceptual movement that is already at work in the text. The main consequence of this thesis is that software implies an interpretation.

In the second section of the chapter, I show why a purely technical definition of software is insufficient. The goal is to show the complexity of software and, therefore, the difficulty of providing an exhaustive definition of it.

In the third and fourth sections, I introduce the central thesis of the chapter: the parallelism between software and text. First, I clarify what Ricoeur means by text, focusing mainly on the three *mimeses* of the narrative function, as analyzed in *Time and Narrative*. In order to facilitate the parallelism with the software, I include numerous references to Don Ihde's post-phenomenology. Secondly, I apply Ricoeur's notion of text to software.

In the fifth section, I apply the Ricoeurian concept of textual "reconfiguration" to software through a concrete example: the logic of Google, the famous "Search, Don't Sort." In the sixth section, I analyze the connections between a hermeneutic approach to software and the so-called Critical Code Studies, which are a completely new field of research. In the seventh section, I try to further expand the discussion on software hermeneutics by analyzing the "narrative turn" in the contemporary philosophy of technology.

This chapter wants to open a bridge between the hermeneutics of software and critical code studies. They share a principle: "The lines of code of a program are not value-neutral and can be analyzed using the theoretical approaches applied to other semiotic systems, in addition to particular interpretive methods developed specifically for the discussion of program [...]" (Marino 2020, 39). Hayles (2005), Manovich (2013), and Kittler (2014)—just to recall some of the most important authors in this field—consider software a literary text to be interpreted. As Marino (2020) points out, code means more than merely what it does; we must also consider what it means. Ricoeur gives us an extra tool to understand how meaning is formed and circulates, and how it transforms the life of the human subjects with whom it comes into contact. As Ricoeur also teaches, hermeneutics makes positive use of literary criticism; it does not have an antagonistic attitude toward it.

What kind of language is software? We could say that it has a performative nature because it does what it says. The set of operations described is immediately performed by the machine. But what relationship is there between that language and the operations, which are simple electrical impulses? What are the boundaries of software? Can we also consider software as the technical documentation that accompanies a programming language, such as Java or C ++? Do only the high-level languages count, or might the compiler's software as well? What is the contribution of users to the "life" of software? There are many different approaches to these questions. In this chapter, I want to explore a new approach: that of philosophical hermeneutics. In particular, I intend to analyze the software as a "text" in Ricoeur's sense. My hypothesis is that software is the text *par excellence*, that is, it fully realizes the concept of text described by philosophical hermeneutics. Therefore, software helps us to better understand the notion of text as well.

A caveat is necessary. Paying attention to the narrative and hermeneutic aspects can lead to neglect of the software considered as a material artifact, moreover as a dynamic and material causal process when executed. For this reason, it is crucial to keep in mind the distinction made in the previous chapter, that between the pure physical level of calculus, the formal level of the mathematical interpretation of the calculus, and the interpretative level of the algorithm. We cannot reduce these levels to each other. The physical level cannot be fully accounted for by mathematical formalization. Mathematical formalization is only formal modeling, that helps design and specify machine behavior and interpret its results.

Ricoeur's approach to narrative is fundamental to understanding the hermeneutic dimension of software and that the most important problem is to conciliate through software a program understood as a narrative and an algorithm understood as a pure calculation without semantics.

## Why We Need a Different Approach to Software

The results of the previous chapter highlighted the complexity of the software and the difficulty of formulating a unique definition of it. In this section, I want to develop this aspect further, showing the limits of an exclusively technical approach to software.

According to Turner (2018),

> Languages and machines represent the two ends of the computational spectrum: the abstract and the physical. They come together at the digital interface, the very lowest level in the computational realm. Digital circuits are employed to store, communicate, and manipulate data. Flip-flops, counters, converters, and memory circuits are common examples. Their building blocks are called gates, the most central of which correspond to arithmetic and Boolean operations. These are simple logic machines, so named because they are intended to represent some form of numerical or Boolean operation. More complex machines are built from them by connecting and composing them in various ways, where the most general-purpose register-transfer logic machine is a computer. (45)

Even though Turner's book remains an essential reference point in computer science philosophy, this view appears to be oversimplified. I contend that Turner does not grasp the complexity of the philosophical problem underlying his premise. From my point of view, the essential software problem is the following: How is the a priori synthesis of logic and machines possible? This synthesis is a priori because it is the condition of possibility of all our digital experiences and is independent of these experiences (the computer also acts in my absence and performs these operations). Moreover, what does the synthesis between the machine, Boolean operations, and user experience guarantee? How can we know that this correspondence takes place?

Boolean logic truth tables give us a functional description of a digital circuit. This description is formulated in a formal language and says how the digital circuit should work. However, the digital circuit is also a physical object, that is, a set of material and electrical pulses. As a physical object, the electrical circuit has a series of structural properties that tell us what it is and what it does (Turner 2018, 33), but the functional and physical properties are incompatible. In fact, the functional properties "provide no account of how the actual electronic devices are to be built; they do not describe how the computations are to be carried out. They are functional specifications, not [physical] ones, and they cannot be easily turned into the latter. They tell us what the actual physical device should do: the what not the how" (Turner 2018, 33). Functional and physical properties are also incompatible with the characteristics of user experience, which incidentally may have knowledge of neither of them.

In Chap. 6 of his book, Turner talks about software ontology. He distinguishes between functional description (specifications), structural description (HL languages), and physical description (implementation). According to him, "in the case of programs, implementation is a mechanism that, given a symbolic program as input, returns a physical process" (49). As computational artifacts, programs mediate between functional and physical properties, yet Turner fails to explain how.

He does not see a problem in this transition from the symbolic to the physical. Software is a complex abstract structure, made up of many levels and languages and levels of abstractions, which is capable of producing a physical effect. It is a language that does what it says. It has a performativity that is independent of any human intervention. How, therefore, is it possible for a symbolic apparatus to produce a physical effect? How can an abstract mathematical structure implement physical operations? Turner limits himself to quoting Colburn's (1999) well-known thesis about "harmony," according to which there is a "fundamental harmony" between the physical and the symbolic in the program. However, this is not an explanation at all—as we saw in Chap. 2. Turner seems to actualize Colburn's solution by saying that, "presumably, it is via the semantics that the programmer is able to design the program from the specification, and it is via the semantics that the programmer is able to explain why and justify the claim that the program meets the specification" and, so, the implementations (Turner 2018, 51). He distinguishes three levels—syntax, semantics, and implementation—and thinks that the passage from the first to the third is due to the second.

Let us consider Turner's (2018, 99–100) example. We have the following formula:

$$A := 13 + 74$$

This string of code can receive a "physical interpretation" that will have the following form: physical memory location A receives the value of physically computing 13 plus 74. However, what does "physical interpretation" mean? In reality, this interpretation is just another formulation of that string of code, that is, its translation into another language. That string of code corresponds to a certain state of the CPU and, therefore, to a series of operations, that is, physical states—what the machine actually does. The correspondence is ensured by the programmer's semantic choices. Nevertheless, as Turner claims, a semantic explanation concerns only the functional level; it cannot tell us anything about the physical operations of the machine. A physical explanation concerns only the structural and physical levels, that is, how the machine effectively acts, and it cannot tell us anything about how the machine should act. Neither the functional nor the structural levels have the resources to explain their connection. In other words, the connection sought can be neither functional nor structural nor physical. Turner (2018, 101) limits himself to saying that "the physical implementation is subject to the abstract interpretation, and the meaning of the construct is given by the abstract account alone." He still does not see the complexity of the problem of the connection of levels or of the relationship between the different levels and user experience.

I make four fundamental criticisms of Turner's approach:

- He does not consider the problem of the interaction, that is, the relation with user experience;
- The basis of his theory is an ontology of the thing and not of the process—for this reason, he defines software as an object, an artifact—but I do not think that this is appropriate to explain software; I think that software must be understood in terms of processes rather than things;
- His is an excessive intellectualization of software. This excessive intellectualization made it impossible to understand important aspects of software such as the artistic pleasure it can produce in those who think and write it, or its narrative dimension.

Let us try to clarify the problem. We have three distinct levels:

The functional level → formal syntax and semantics
The physical level/1 → implementation/problem-solving
The physical level/2 → user experience

Neither a monistic solution (all levels can be reduced to one; they form a unitary whole) nor a dualistic solution (the physical and abstract are two distinct and non-communicating levels) appears satisfactory. If we choose the first solution, we have to explain the differences between these levels. If we choose the second way, we must explain the unity of these levels. Turner chooses the second way, which is the reason he is forced to invoke an inexplicable "harmony." He also refers to Landin's correspondence principle (Turner 2018, 170–71), but this changes nothing. Separating the functional level from the physical level is only an intellectual abstraction. All the levels are connected in the digital experience: user experience is a continuum.

## Ricoeur's Notion of Text

I propose to introduce the parallelism between software and text by referring to the post-phenomenological notion of "multistability." Post-phenomenology is obviously only one possible approach to the philosophy of technology. However, I think it is a very useful approach for the purposes of this chapter.

Post-phenomenology is "a particular mode of science-technology interpretation," Ihde writes (Rosenberg and Verbeek 2017, 1). This approach combines the style of Husserl's phenomenological investigation and the tradition of American pragmatism. Science and technology are conceived as a fundamental mediation between the human being and the world. They shape our way of relating to and thinking of the world. For example, in Ihde, the main exponent of post-phenomenology, the relationship between technology and the human being is conceived in four ways: embodiment relations (when a technology is "embodied," a user's experience is reshaped through the device), hermeneutic relations (the use of a device involves an interpretation), alterity relations (the devices to which we relate in a manner somewhat similar to how we interact with other human beings),[1] and "background relations" (the devices that define the environment in which the subject lives). Therefore, technologies have to be understood in terms of the relations human beings have with them, not as entities "in themselves."

By focusing on mediation, post-phenomenology reconceptualizes the intentional relation[2] in two distinct ways. "First, [post-phenomenology] investigates its fundamentally mediated character. There is no direct relation between subject and object, but only an 'indirect' one, and technologies often function as mediators. The human-world relation typically is a human-technology-world relation" (Rosenberg and Verbeek 2017, 12). Secondly, post-phenomenology "does away with the idea that there is a pre-given subject in a pre-given world of objects, with a mediating entity between them. Rather, the mediation is the *source* of the specific shape that human subjectivity and the objectivity of the world can take in this specific situation. Subject and object are *constituted* in their mediated relation" (Rosenberg and Verbeek 2017, 12). Hence, intentionality is not a bridge between subject and object, but rather a fountain from which the two of them emerge. Starting from this rereading of intentionality, post-phenomenology develops a relational ontology.

However, how should we understand the way that technology at once in part determines our choices and actions, and yet at the same time itself remains open to our manipulation and interpretation?

In order to answer this question, Ihde introduces the concept of "multistability." This concept refers to the idea that any technology can be put to multiple uses and can be meaningful in different ways to different users. As Ihde puts it, "No technology is 'one thing,' nor is it incapable of belonging to multiple contexts" (Ihde 1999, 47). This means that a technology can have many stabilities, that is, defined uses, in a number of different, variable contexts. It does not mean that a technology can mean anything. A technology adapts to many different contexts, creating different types of possible experiences. As Ihde writes, "a hammer is 'designed' to do certain things, but the design cannot prevent the hammer from (a) becoming an object d'art, (b) a murder weapon, (c) a paperweight, etc. Heidegger's insight was to have seen that an instrument is what it does, and this in a context of assignments" (Ihde 1999, 46–47). The concept of multistability tells us that every technology requires an interpretation in relation to the context in which it is used.

The issue I want to examine is the following: Can we apply the notion of multistability to software? Does software require interpretation? Which kind of interpretation? Multistability can be understood as the property to have a transcendental value (it is an a priori condition of human experience) and an empirical value (it is built by human beings). I will clarify more about this transcendental dimension of technology and software in Chap. 5, when I talk about Digital Reflective Judgment.

Software, like any technology, is a form of mediation: it allows us to use a machine (the computer) to perform some operations and achieve some tasks. Hence, following Ihde, software also requires an interpretative act. My guess is that Ricoeur's concept of text is a very powerful tool that can help us better understand the nature of the interpretative process taking place in software. In philosophy of technology, David Kaplan (2006) has claimed that we use narrative to make sense of technologies and has suggested that the work of Ricoeur offers "a narrative theory of interpretation for making sense of all the different ways that technologies figure into our lives" (Kaplan 2006, 50).

The reason we choose to be guided by Ricoeur is above all methodological. Ricoeur's thought seems to be a particularly versatile philosophy due to his unsystematic character. Even though Ricoeur did not construct a closed philosophical system, there is a coherence in his thought and the way in which he approaches his concern with philosophical anthropology. The unity of his thought is analogical, flexible. Ricoeur has traced the way in which language mediates one's understanding of oneself and one's lifeworld through different stages of linguistic acts. For this reason, Ricoeur's philosophy appears not only as a philosophy of mediation, which always tries to mediate between pairs of conceptual extremes, but also as a mediating philosophy, capable of making very different philosophical traditions dialogue. In his writings about language, Ricoeur elaborates ideas from both the Anglo-American tradition ("analytic") and the European tradition (commonly referred to as "continental"). This pushes him to go as far as to compare the works of two philosophers who at face value might be considered each other's opposites: Husserl and Wittgenstein. Analytic philosophy of language (Austin's and Searle's speech act theory and Davidson's agentless semantics of action) significantly influenced Ricoeur. At the same time, his early philosophical training was based above all on continental philosophy, in particular the phenomenology of Husserl and Heidegger, which he tried to develop in an original way. Equally important is the contribution of Gadamer's hermeneutics, the semiotics of Greimas, and the historiography of Hayden White and Marc Bloch.

Essential to Ricoeur's understanding of language is the interpretation of text, which he argues can be seen as a model for meaningful action. The core concept of Ricoeur's hermeneutics of text is that of distancing, or "distanciation." Ricoeur defines the text as "written discourse." Through writing, the language becomes autonomous from (1) the intention of the original author, (2) the original world of circumstances in which the author wrote or which s/he wrote about, and (3) the original readers of the text when it was first produced (for instance, the Greek community who read Homer's *Odyssey*).

However, in this process, "distanciation" is connected to another process that goes in the opposite direction. This process is called "belonging," or, following Gadamer's terminology, the "appropriation." The autonomy of the text is the condition for the text to be read and understood by several readers, who re-read their own experiences through the text and transform its meaning. From this point of view, following Heidegger and Gadamer, Ricoeur talks of a dialectics between distanciation and appropriation. He overcomes the structuralist point of view on text and literature by arguing that the discourse can never be completely reduced to its syntactical and grammatical structures. Structuralism's main thesis is correct: texts have an internal linguistic structure. However, this internal structure projects itself outside.

For Ricoeur, the language "goes beyond itself" because it is essentially mediation between the subject and herself/himself, between the subject and other subjects, and between the subject and the world. Therefore, language is not only a set of symbolic structures "internal" to the text but also a movement that refers to the external world and the reader's praxis (see Ricoeur 1965, 1969, 1986). Thanks to its structures, the text "projects a world," says Ricoeur, in which the subject recognizes himself and his way of being. Imagination plays an essential role in this process. The imaginative work of the reader responds to the imaginative work of the text. The text is "a heuristic model" given to the reader, that is, a tool that allows the reader to discover new aspects of her/his experience and praxis. Imagination makes the reader able to translate what the text tells him/her in his/her existence and praxis. As Kearney (2006, 16) claims, Ricoeur "argue[s] that the meaning of Being is always mediated through an endless process of interpretations—cultural, religious, political, historical and scientific. Hence Ricoeur's basic definition of hermeneutics is the art of deciphering indirect meaning."

The dialectic of the text is at the core of *Time and Narrative*; this is the paradigm of the text (see Ricoeur 1983–1985). Ricoeur's theory of narrative "revolves around a basic model that designates the way in which a narrative, considered according to the paradigm of the text, mediates human reality" (Coeckelbergh and Reijers 2020, 81). In *Time and Narrative* Ricoeur distinguishes three moments of the text's "life": the pre-understanding of the world of action (its meaningful structures, its symbolic resources, and its temporal character); the configuration of the text, that is, the emplotment, which is the production of what Ricoeur calls "productive imagination" (expression drawn from Kant); and the intersection of the world of the text and the world of the hearer or reader, that is, the set of effects of the text on concrete experience. The text is not an object, but a three-step process. These three moments do not form a circle, but a spiral, a movement that returns to itself, but always in a new way. "That the analysis is circular is indisputable. But that the circle is a vicious one can be refuted. In this regard, I would rather speak of an endless spiral that would carry the mediation past the same point a number of times, but at different altitudes" (Ricoeur 1984, 72).

Let us take a closer look at these three stages. First, for Ricoeur, the fundamental model of the text is the narrative. What is narrative?

> Plainly said, deriving from the Aristotelian tradition, a narrative like a story in a book is defined by the way in which it mediates the experience and understanding of the reader's social world. This mediating role can be captured by the concept of emplotment, which designates the capacity of a narrative to organize heterogeneous elements (characters and events) in a meaningful synthesis that we commonly designate as the plot. As such, we characterize narrative as a mediation of human experience and understanding of the social world through the process of emplotment, which is the organization of heterogeneous elements in a meaningful synthesis. (Coeckelbergh and Reijers 2020, 42)

According to Aristotle, and Ricoeur takes up this idea, narrative is the imitation of action. Consequently, to understand any narrative text a preliminary competence is necessary: "the capacity for identifying action in general by means of its structural features" (Ricoeur 1984, 54). Aristotle says about mimesis that "the instinct of imitation is implanted in man from childhood" and that this is an "instinct for 'harmony' and rhythm" (Aristotle 1902, 15). I cannot understand the imitation (or representation) of an action if I do not already know what an action is, and I cannot recognize it. This understanding is pre-linguistic and pre-logical; it has a practical nature. This condition can also be extended to the understanding of non-narrative texts: there is an essential pre-understanding—a common sense, I would say—that allows us to follow a text and learn its meanings. Between this necessary practical pre-understanding and the text, there exists "a relation of presupposition and of transformation" (Ricoeur 1984, 55).

The second phase is described by Ricoeur as a synthesis of heterogeneous elements. In the case of the story, "this configurational act consists of 'grasping together' the detailed actions or what I have called the story's incidents. It draws from the manifold of events the unity of one temporal whole" (Ricoeur 1984, 66). This synthetic act is very similar to the Kantian reflexive judgment; "[…] the kinship is greater still with the reflective judgement which Kant opposes to the determining one, in the sense that it reflects upon the work of thinking at work in the aesthetic judgment of taste and in the teleological judgement applied to organic wholes" (Ricoeur 1984, 66). The rules on which this synthesis is based are produced by the imagination through a twofold process of sedimentation and innovation (Ricoeur 1984, 68–70). This second phase is the time of structure and objectivation; the text becomes an object thanks to the writing and internal structure. This is the construction of the plot. "The plot is the imitation of the action—by the plot I mean here the arrangement of incidents" (Aristotle 1902, 25). From this, Ricoeur affirms that emplotment designates the organization of events by which people represent action in a plot. "Paradigmatic examples are works in the genres of tragedy and comedy in which characters imitate probable accounts of human actions structured according to a play script or scenario, which is an organization of events (or 'acts')" (Coeckelbergh and Reijers 2020, 82). In a narrative such as a tragedy, the plot configures different elements like characters, motivations, and events in a meaningful whole. "Emplotment, in other words, creates a harmonious concordance out of discordant, heterogeneous elements" (Coeckelbergh and Reijers 2020, 82). The synthetic unity of the plot gives a sense to the events; it is the creation of a meaning, of a teleological structure. This unity is the object of the act of reading, and it is also the principle of the reader's understanding. The perception of this unity leads the reader to affirm that "this story makes sense."

The third phase is the reference, that is, the meeting between text and reader. This is the moment at which the narrative circle is completed, and the lifeworld of the reader is transformed. The text modifies our practical pre-understanding of the world in the sense that it enriches and transforms it. The text "is a set of instructions that the individual reader or the reading public executes in a passive or creative way" (Ricoeur 1984, 77). The text offers the reader a set of possibilities. These possibilities are of two types: (a) the possibility of analyzing its experience in a new way, and above all our pre-understanding of the world; (b) the possibility of acting in a new way, that is, of enriching its way of acting in the world together with others.

In the third phase of mimesis, which corresponds to the act of reading, the emplotment becomes a joint creation of the text and the reader. Ricoeur explains it in this passage:

> It is the act of reading that accompanies the narrative's configuration and actualizes its capacity to be followed. To follow a story is to actualize it by reading it. And if emplotment can be described as an act of judgment and of the productive imagination, it is so insofar as this act is the joint work of the text and reader, just as Aristotle said that sensation is the common work of sensing and what is sensed. (Ricoeur 1984, 76)

We can summarize the three phases of the text's "life" through Fig. 3.1:

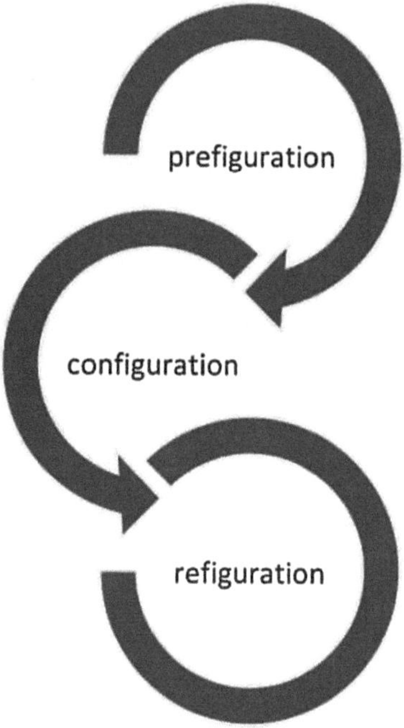


**Fig. 3.1** The three phases of the "life" of the text according to Ricoeur's *Time and Narrative*

## Software as Text

To introduce the application of the Ricoeurian hermeneutic model to software, which is the central point of the chapter, I propose a general distinction. Following Primiero (2020), I distinguish between ontological domain and epistemological structures (see Table 3.1):

**Table 3.1** Ontological domain and epistemological structures in a computational system

| | Ontological domain | Epistemological structures |
|---|---|---|
| 1 | Intention | Problem |
| 2 | Algorithm | Task |
| 3 | Programming language | Instruction |
| 4 | Machine code | Operation |
| 5 | Electrical charge | Action |

The scheme (taken from Primiero 2020; Fresco and Primiero 2013) is simple in itself. Each level provides a different type of data. "Each layer in this ontological domain is associated with the one above it: an electrical charge is controlled by machine code, which is denoted by a programming language construct, which implements an algorithm, which satisfies an intention. The explanation of the ontology requires an appropriate epistemological structure [...]. Each epistemological construct has a relation with the underlying one: a problem is reflected by a task, which is interpreted by an instruction, satisfied by an operation and executing an action" (Primiero 2020, 174). Software is that level allowing the passage from the top to the bottom of both schemes. It is therefore a mediation between a first rationality (levels 1–2) that identifies a problem and establishes the characteristics (specifications) that its possible solution must have, thus defining a process with a certain purpose, and a second rationality (levels 4–5) which translates this process into operations that can be performed by a physical machine. These two rationalities have a practical nature: the goal is to define a concrete action solving a problem, and therefore fulfilling a purpose. Software translates problem-solving into machine actions. This explains the dual nature of software, which must mediate between a more abstract and a more concrete rationality. The nature of this mediation is a problem though—literature talks about implementation, "the problem of explaining the link between the abstract and physical layers of computation" (Primiero 2020, 190).

Primiero's scheme can be improved. Between the intention and the algorithm there is an intermediate level: modelization. Modelization is (1) theoretical and (2) mathematical. Only after a mathematical representation of the theoretical model, one can define the algorithm.

Now, Ricoeur's hermeneutical model can be a powerful tool to understand the dynamic nature of software.

First, software can be interpreted as a higher degree of "distanciation." In software, a language (a set of characters and rules) becomes autonomous with respect to its author (programmers and designers), the circumstances, and the original audience, as Ricoeur's notion of text. Software maximizes the autonomy of writing. A program uses the autonomy of writing in order to act in the world. By software, writing becomes as social an agent as any human being. By software, writing becomes so

**Table 3.2** The four levels of software

| | Software |
|---|---|
| 1 | Machine code (binary) |
| 2 | Compiler |
| 3 | Programming language |
| 4 | Documentation |

autonomous that it is no longer meant to be read—I mean that its main objective is not to be read. As Chun (2013, 91) says, "for a computer, to read is to write elsewhere." Software is thus the realm of pure writing. "Software's specificity as a technology resides precisely in the relationship it holds with writing, or, to be more precise, with historically specific forms of writing" (Frabetti 2015, 68). "Software is a special kind of text, and the production of software is a special kind of writing" (Sack 2019, 35). Software takes to the extreme the nature of writing.

Writing in software develops on several layers. Some layers are material, others immaterial, some visible, others invisible, some imply the participation of the human being, and others do not. I propose an overview in the following chart (see Table 3.2):

Software is a writing and re-writing process that goes through these four levels.

A movement of increasing abstraction takes place through these levels. The language becomes less and less human and closer to the machine, until it becomes pure machine code, that is, binary language. The binary code is understood by the machine and executed. Through these levels a translation process takes place: the user gives a command, and this command is defined by an interface and then expressed in HLL (high-level language). The code string in HLL is translated into a compiler, that is, in another language, which completely restructures the code to make another translation possible, the one in machine code. These steps are purely "internal," syntactic, and structural, as computer science calls them.

There are two intermediate levels between the HLL and the machine core, the CPU. The first one is the operating system, which "forms the host machine on which a high-level programming language is implemented" (Gabbrielli and Martini 2010, 22). The second is the set of languages that translate (or re-write) HLL into machine code. Indeed, the machine code must be a binary language code because the CPU can understand just strings of 1 and 0. Therefore HLL must be "translated" into machine code to be implemented. Then the machine code is translated into electrical voltages (Elahi 2018, 161).

Re-writing HLL into machine code is the task of another language, the compiler, which restructures and re-writes the entire program. The term "compiler" means "a program that translates instructions written in a language that human beings are comfortable with, into binary codes that a computer can execute" (Ceruzzi 2003, 85). Through a complex procedure of lexical, syntactic, and semantic analysis, the compiler produces an intermediate code, the assembly, which is subjected to an optimization process. The assembly is called a low-level language.[3] It is not an exact translation: the compiler integrates the program by modifying it—for example, by identifying errors or making interpolations. The compiler/assembler allows the translation of HLL into machine code. These strings correspond to the operations to be performed by the CPU. At this point, software is actually realized, and it performs its function and acts in the world. "Each of these instructions tells the computer to undertake a simple task, whether to move a certain piece of data from A to B in the memory, or to add one number to another. This is the simplest processing level of the machine, and it is remarkable that on such simple foundations complex computer systems can be built to operate at the level of our everyday lives" (Berry 2011, 96).

Thanks to this series of writing and re-writing levels that correspond to the level of configuration (second phase in Ricoeur's scheme), software allows the passage from the first rationality of the algorithmic problem solving (problem-task) to the second rationality of the concrete action of the machine (operations-actions). This passage can be read in hermeneutic terms, that is, as the passage from the prefiguration of the action to the reconfiguration of the action. The difference from Ricoeur's process is that here the reconfiguration process takes place in two places: the machine and the user. In this way, software not only confirms the Ricoeurian model, but even amplifies it.

At the root of the work of the programmer—and of the programmers in a group—is what Ricoeur calls the "configurational act," which, as mentioned above, has the same form as the Kantian reflective judgment. This type of judgment is, for Kant, an act of the mind that "reflects on" a series of objects already known and gives them unity and coherence. In software, just as in the narrative, the configurational act has to cope with instructions and operations. The problem-solving scheme that the designer has in mind (problem-task) is modelized and translated into a series of instructions according to (a) the rules of the chosen programming language and (b) the personal style of the programmer. I will analyze more the structure of the configurational act in software in Chap. 5.

How does the translation of the problem task into software take place? Following Ricoeur, we claim that the imagination allows this operation. Ricoeur (1978) describes the imagination as the "ability to restructure semantic fields." This means that, in our case, the imagination restructures the semantic field of action (the set of meanings by which we identify and understand the action) and language (the way we use and understand language). First, the imagination restructures the field of action, allowing it to be rethought in algorithmic terms, that is, according to the scheme of an algorithm and in relation to the needs of the problem. Second, the imagination restructures the language by allowing the translation of the algorithm into instructions for a machine. The concept of action is radically restructured; it is split by the human being. The agency no longer belongs only to humans.

This double restructuring work leads us to the third phase of the scheme: the reconfiguration, that is, the second rationality. Just as the story reaches its full meaning only when it is returned to concrete human life—"the time of acting and suffering," as Ricoeur says—software also acquires full meaning only when it is translated into the compiler and therefore into the machine code in order to become a series of machine actions. These actions have effects and produce ever new meanings.

According to Ricoeur, the reconfiguration occurs in reading. Reader and text meet and shape each other. The text offers the reader a model of action to which the reader is called to respond. In a nutshell, we use narrative in order to explore new ways of acting in the world. The meaning is born from the encounter between the reader and the text, which is always an actualization of the text.

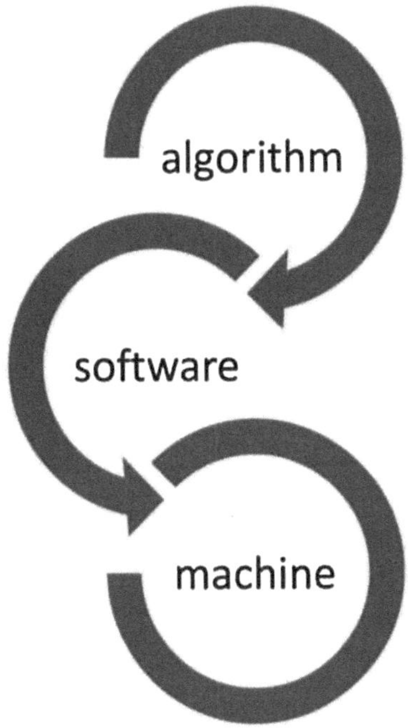


**Fig. 3.2** The three phases of the "life" of software

In software the same happens; machine and human user interact, and this allows a transformation of the human action. The programming language is translated into the machine code which is then translated into electrical impulses and into actions in the human world. This is a hermeneutical circle. Software distances itself from the world, but only to return to the world and transform the concrete experience of the subject, that is, the user. This is summarized in Fig. 3.2

The reference to post-phenomenology, and above all to Ihde's work, can help us in better understanding the concept of reconfiguration applied to software. Ihde (1990) and Verbeek (2011) distinguish at least six types of technologically mediated human-world relations:

1. **Embodied relations**: technology helps to develop and improve the perception and use of the body—for example, glasses;

2. **Hermeneutic relations**: technology requires an act of interpretation to be used—for example, the display on a screen;
3. **Alterity relations**: technology suspends our relationship with the world and becomes the main interlocutor of the human subject—for example, the case of the video game;
4. **Background relations**: technology defines the context of the human subject, that is, a set of tools with which the subject does not directly interact but which define and influence his/her environment—for example, electric light, and heating.
5. **Cyborg relations**: the border between human subject and technology is very blurred—for example, the use of psychotropic drugs and neural implants;

6. **Immersion relations**: technology does not mix with the human subject, but with the natural environment—for example, cities and plantations.

Now, my thesis is that software introduces a new form of relationship between human being, technology, and the world that redefines all these relationships. For example, software (and the digital in general) allows us to mix and join different relationships, or to strengthen them. This is the reconfiguration in software. This is tantamount to claiming that the meaning of software is determined not only by programmers, but also by how it is received and recirculated (Marino 2020, 4).

It can be summarized by the following scheme:

$$\text{software} \rightarrow \left(\text{human subject} - \text{technology} - \text{world}\right)$$

With software, writing (a form of technology) "enters" into things and processes, and modifies the world. It is a new form of writing that exploits the original autonomy of writing to control and modify things and processes.

The code is the new "symbol that gives rise to thought" (Ricoeur 1959). Just as the narrative transforms the subject's relationship with himself/herself, with others, and with the world, software does exactly the same by interacting with the world. Turkle's studies on the psychological impact of new digital technologies clearly prove this (Turkle 1984). However, the effects are not only psychological but also economic and social. Our era is that of the "algorithmic war" (Amoore 2009), in which companies, institutions, and governments struggle daily to obtain and manage our data for security, commercial, scientific, or political propaganda purposes. Software changes the inner nature of capitalism. We live in the age of "knowing capitalism" (Thrift 2005), that is, capitalism that thrives on data because information generates profit. The data-information-capitalism circle produces new forms of social organization, power, and inequities (Baldwin 2016). In this situation, software is the main asset. Yet this involves new risks. Knowing capitalism is a hyper-controlled form of society and economy in which privacy and security are jeopardized constantly. In short, "technological innovation will very likely continue to open new paths into the global economy and continue to spawn uneven reversals and countertrends such as Brexit in the UK, Trumpism in the US, and the rise of right-wing populism in Europe" (Elliott 2018, 70).

However, there is an even deeper transformation that digital technology has imposed on our way of thinking and knowing. This is a revolution triggered by large AI software systems (machine learning) and so-called Big Data. To illustrate it, I use an expression from Carpo (2017): The Second Digital Turn.

## The Second Digital Turn

Data science is a very important topic today and questions raised by these new approaches address the problem of data as traces, records, that is, writing the reality. Re-writing traces by abstracted laws or by induced correlations are two competing paradigms in AI and data science. This is an effect of the use of software on our systems.

Carpo (2017) distinguishes two different logics in data science: the search and the sort. The sort is the classic method of science, which has two essential characteristics: the ability to analyze small amounts of data and abstraction. These characteristics are deeply connected: it is because we are able to analyze just a limited amount of data that we need general formulas and abstract schemes capable of predicting and explaining data that we do not know. The formula applies to the individual case in order to explain it, that is, connect it to a causal network.

The search logic is completely different. It has two essential characteristics: the access to a potentially infinite amount of data and the use of machines that allow us to analyze this data at great speed. Search is a logic of design, not of causal explanation. The search does not use general formulas, that is, abstractions, because it does not need to compress the data, to reduce them to recursive structures. It does not need to reduce the individual case to a universal category and a causal link. Search is the logic of Google, the philosophy of "Search, Don't Sort."

Let us take the example of a cyclone. The classical scientists act more or less in the following way: they analyze the data in their possession and reduce it by means of a law, that is, a set of mathematical formulas, that is a model. They carry out a series of experiments on cyclones and collect a limited quantity of data. Starting from these data, they extract patterns, that is, recurring structures that they call laws and are formalized in mathematical structures. Therefore, in order to study the behavior of a single cyclone, they simply apply these formulas to the data in their possession.

The search scientist does not need to compress data through formulas. Thanks to digital technology, they extract all the possible data concerning the single cyclone they have to study. Using this almost infinite quantity of data, they propose an answer to a given problem. At this point, the simulation intervenes: using a very powerful computer, the search scientist tests the answer. At the end of this simulation, they will have an almost infinite number of answers to problems connected to that type of situation. The search logic proposes an alternative to modeling.

As I mentioned above, the search approach is the Google approach. Google does not sort; it does not use categories. Google searches based on criteria we give it. Humans do not search, but sort. Being able to analyze only a few data at a time, humans need to sort in order to know where and how to search. Faced with a very long list, the human being looks for recurring patterns that allow them to compress the data and therefore know the structure of the list and understand an order. Google does not do this: it analyzes every element of the list at a very high speed until it finds what it is looking for.

The sort was brought to its extreme by Stephen Wolfram in his controversial book *A New Kind of Science* (2002). In this book of over a thousand pages, Wolfram shows that through the use of cellular automata (which he calls "simple programs," that is, very simple algorithms), a computer can simulate events that mathematics and physics cannot foresee or even conceive. Wolfram shows that the iteration of very simple operations in these cellular automata is trivial and empty only to human eyes. In fact, huge amounts of iterations of simple operations can be creative and produce something—that is, complexity. By the use of cellular automata, we can simulate chaotic and complex events that traditional science has difficulty explaining—however, it should be said that Wolfram's theory has been highly criticized. The search approach is different because it proposes an alternative to the sort.

A different important approach to data science is that of Pearl and Mackenzie (2018). "Data are profoundly dumb. Data can tell you that the people who took a medicine recovered faster than those who did not take it, but they can't tell you why" (Pearl and Mackenzie 2018, 6). We need causal models that do not come from data but from human intuition. Machine learning programs just capture associations; they do not go beyond this. Beyond data and their associations, there is causality. Causality cannot be reduced to probability; "the idea of causes and effects is much more fundamental than the idea of probability" (Pearl and Mackenzie 2018, 46). We need a specific form of calculus in order to understand causality. Pearl's calculus combines statistical mathematics with the language of causality—in other words, it offers a causal interpretation of statistical rationality. It is based on two elements: causal diagrams and do-calculus. The former are simple diagrams composed of arrows and nodes depicting relations and data, while the latter is a mathematical language based on three laws. The crucial distinction is that between seeing and doing if we stop at seeing—at the mere observation of the data in our possession—we cannot grasp the doing, that is, the intervention, the cause-effect relationship in the set of variables (see the concept of "Ladder of Causation" in Pearl and Mackenzie 2018, 28). The do-calculus allows us to distinguish association and causality, and to precisely identify the nature of the causal link between variables. Furthermore, the do-calculus also allows us to explain the counterfactuals.

This is a very important point. If we follow the indications of Pearl and Mackenzie (2018), we can claim that the narrative is an essential tool in data analysis. The narrative configuration analyzes causal relationships. The plot is a set of causal relationships—or at least, it is the author's effort to distinguish the real causal relationships and the mere associations. A narrative science of data should be based on the Ricoeurian hermeneutic paradigm. The refiguration is also a form of causality, in the sense that the tale has an effect in the real world, in the redefinition of the reader's life. Narrative is a third paradigm beyond search and sort.

## Critical Code Studies

We have talked so far about the practical effects of software on human reality. But is it possible to read the code, by which I mean the *source code*? What does it mean to interpret software? Reading and critically analyzing program code is not like reading a novel. Software has a completely different structure from any other text.

The hermeneutic approach of the present book is very close to critical code studies. Critical code studies "names the applications of hermeneutics to the interpretation of the extrafunctional significance of computer source code" (Marino 2020, 33). Indeed, the thesis of critical code studies is that "in the process of its circulation, the meaning of code changes beyond its functional role to include connotations and implications, opening to interpretation and inference, as well as misinterpretation and reappropriation" (Marino 2020, 4–5). Code is "a social text, the meaning of which develops and transforms as additional readers encounter it over time and as contexts change" (Marino 2020, 5). Software is a text and has a personal style that can be interpreted in many ways. The same function can be written in many different ways; the style of writing reveals an identity, a course of study, an aesthetic choice, a way of communicating. The meaning of the code goes beyond the function, but at the same time is deeply connected to this latter. "Code holds a meaning beyond, and growing out of, what it does" (Marino 2020, 9).

Critical code studies are a family of reading methods and practices whose constant evolution is strictly connected to new technologies. They explore "the metaphors raised by the names of constructs in the code, examining misreadings and recontextualizations of code, interrogating the cultural stakes of language formation, and reading code as a complement to philosophical tracts" (Marino 2020, 227). Although still in their infancy, critical code studies could become the future of Humanities. If, as very likely, in the future programming languages will become the most widespread languages on planet Earth, then the ability to interpret these languages, and therefore of a new type of literary criticism, will be a crucial element in the education of any person (and probably any machine too). As Montfort (2016) explains, programming is not only a mere technical exercise but an experimental activity that is culturally situated. For this reason, understanding the functions of the software is only the first step in reading and interpreting the software. Beyond each code, there is the story of its creation, "the concept behind its conception, the context of its development, and the constraints face by its creators" (Marino 2020, 233). The code embeds also metaphors and symbols that illustrate the imaginative work behind its creation, and therefore also the cultural context in which it arose. Marino highlights this aspect above all in the interpretation of Kittler's Raytracer code, whose central metaphor is the couple Heaven and Hell (Marino 2020, 177–178).

Of all the authors who can be linked to critical code studies (Manovich 2013; Hayles 2005), Marino is the one who elaborates the most accurate methodology. In fact, interpreting a code can be a very complex undertaking. Marino proposes several general techniques (2020, 27–28; I gave just some examples):

- Reading the code and its documentation to determine what the code does;
- Analyzing the embedded metaphors and imaginary constructions (Marino 2020, 80);
- Discussing the code with its authors;

- Scanning the overall architecture while reading comments if available; guide and commentary texts can be useful;
- Using additional software to see the code in action and monitoring changes;
- Research on the programming language;
- Research on any required hardware and software;
- Research on the evolution of the code.

The great difficulty of reading and interpreting the software is due to its intrinsic complexity, due to several factors:

- The lines of code can be in the millions.
- The authors of the code can be many and different (humans and machines); software puts the classic role of the author in question (Foucault 1982).
- Software receptors can be many and different (humans and machines).
- Software is not static; the code reader has to comment the code in action, which is very complex due to the intricacy of the interaction between the running code and the system (or systems).
- Software may work differently on different machines or on different parts of the same machine.
- There are many different approaches to software (platform studies, media archeology, semiotics, etc.).

An education is also located; there is someone who thinks it, designs it, and elaborates it, and someone else who translates and realizes it. Both of these actors are situated because they belong to a culture, a history, a reference context. Even computation is shaped by historical and cultural processes (computation is also rhetoric; see Golumbia 2009). Even the choice of a programming language is the expression of an ethos, of an affiliation, of a series of values that distinguish a cultural and social belonging.

The task of critical code studies is to unpack the hidden network of meanings in the socio-technological context. The task of a philosophical hermeneutics of software is to develop an existential analysis starting from the critical study of software. What is the existence of humans mediated by software? As Ricoeur points out, ontology is always connected to language, which is the origin of meaning for humans. What kind of ontology is, then, a software-based ontology? Critical code studies explore the software structures and models from which hermeneutics must build—through a further act of ontological interpretation—the new ontology of the digital world.

## Software and the "Narrative Turn"

In this chapter, I have tried to show how software can be understood as a hermeneutic process capable of re-describing the human experience. However, one aspect must be better highlighted: how software re-describes the human experience not only on the perceptual level, but also, and perhaps above all, on the narrative level. Software defines practices and conveys new ways of narrating existence. It does this not simply through "what it does," but also through "what it is," that is, as a cultural object.

How is it possible to connect storytelling and technology? "It is in the narrative mode that we explain our actions-with-technologies; that their 'agency' is revealed. It is in the narrative mode that technical practices gain significance" (Coeckelbergh and Reijers 2020, 4). In technology, practices and narratives are intertwined, so that practices, that is, families of actions, give rise to narratives and symbols, which at the same time give rise to practices. The rise of a new technology is always accompanied by new practices and narratives: "It is in narratives that we find the clearest "reflection" of technical practices; the topos where we can read how technologies mediate our actions" (Coeckelbergh and Reijers 2020, 4). Understanding technology requires the particular form of explanation that is narration. Two basic examples are the image that Silicon Valley has been able to build of itself and the almost mythological narratives of the "great founders" of big tech, first and foremost Steve Jobs. These narratives are an integral part of the technologies produced by Apple and other major tech companies.

It seems interesting to note that connecting technology and narration also means questioning the materialist paradigm of classical post-phenomenology (Ihde, Verbeek), which places the materiality of technologies at its center. As Coeckelbergh and Reijers (2020) underline, overcoming the dichotomy between materiality and sign means conceiving technological artifacts as texts, metaphors, and symbols capable of configuring and reshaping human experience in narrative form. The Ricoeurian hermeneutic model, which is these authors' point of reference, represents a middle way between materialism and idealism, providing the possibility of analyzing the empirical turn's weak points (Achterhuis 2001). It is indeed true that "in its execution of the empirical turn, post-phenomenology forfeits a phenomenological dimension of questioning" (Zwier et al. 2016, 1).

The contemporary philosophy of technology has taken an "engineering" approach that starts with an analysis of the nature of technology itself (Mitcham 1994, 62), an approach inspired by Latour's Actor-Network Theory. The rejection of the Heideggerian conception of technology and the attempt to develop a new point of view on technology starting from the concept of mediation is evident in the work of Ihde and Verbeek, as we have seen above. Following Ihde, Verbeek (2005, 2008) has rejected Heidegger's pessimistic and deterministic views on technology at large (rather than on concrete technologies). Furthermore, Ihde (1999) has not only rejected Heidegger's philosophy of technology but also distanced his perspective from Ricoeur's, transforming textual hermeneutics into a material hermeneutics. However, in its critique of Heidegger, post-phenomenology implicitly demonstrates that it adheres to what Heidegger defines as the essence of technology, namely *Enframing*. Post-phenomenology—due to the empirical turn—is itself technologically mediated and therefore fails to truly grasp the essence of technology. To accomplish that, "a rehabilitation of the ontological dimension in the questioning of technology" is necessary (Zwier et al. 2016, 329). In other words, classical post-phenomenology falls victim to its critique of Heidegger.

Other authors, in particular Reijers and Coeckelbergh (2020), highlight three important gaps in classical post-phenomenology that make the definition of a new research paradigm urgent. The first gap concerns language: due to the focus on material artifacts, there has been insufficient attention paid to the role of language, including narrative. "Technological artefacts are assumed to mediate the relation between us and the world, but without any role assigned to language; the words we use and the narratives we recount to ground our understanding of technology" (Coeckelbergh and Reijers 2020, 28). Language is a form of technology that implies a specific mediation and form of interpretation: "Consider an android robot that evokes fear as a response: to understand this, we have to refer to prefigured understanding, shaped by narratives about technology such as Mary Shelley's *Frankenstein*, a story which is usually interpreted as a warning for more intelligent and human-like technology running out of control (or, a story about a human creator abandoning his creation)" (Coeckelbergh and Reijers 2020, 28). But language is not just narration. It also possesses a profound symbolic dimension, as demonstrated by the history of symbolism and metaphors. This is an interpretative and imaginative dimension that escapes the method of classical post-phenomenology more focused on the materiality of technologies.

The second gap concerns time, the temporal dimension experienced by a human being. "The postphenomenological analysis is indifferent regarding time: it does not consider that it is made at one point in time—a momentary intervention of technology—and human subjectivity in relation to the technology is understood in a rather a-temporal way" (Coeckelbergh and Reijers 2020, 29). Even the Critical Theory of Technology does not consider how the human experience of time affects technology (Coeckelbergh 2020, 91). Everything seems flattened into an eternal present without evolution and without perspective, in which the technological object appears somewhat magically out of nowhere. This loss of the dimension of time is one of the consequences of the rebellion against Heidegger.

The third gap, on the other hand, concerns the subjectivist dimension of post-phenomenology. When describing the different forms of human-technology relations, Ihde always refers to the subject as "I." This is a legacy of Husserl's phenomenology. In contrast, there is no attention given to the social dimension of technological mediation. "The social aspect of how technology mediates and shapes our experience and action is under-theorized, if not absent. Ihde's postphenomenology does not pay much attention to the social and instead focuses on how technology mediates human-world relations" (Coeckelbergh and Reijers 2020, 29).

In order to fill these three gaps, many philosophers of technology propose a new form of post-phenomenology that holds together the notions of language, time, society, and materiality. This is a more complex and varied paradigm. In this framework, the concepts of narrative and practice are used as conceptual tools to create a synthesis between the four key notions. The crucial idea is that practices—intended as "any coherent and complex form of socially established cooperative human activity through which goods internal to that form of activity are realized" (MacIntyre 2007, 187)—and narratives define each other. Coeckelbergh (2017) has moved in this direction. Inspired by Wittgenstein, he proposes the notions of "technological game" and "techno-performance" to demonstrate how the technological object is inserted into a wider and language-mediated practice. However, in this approach, narrative does not really have a place as an essential technological mediation. Coeckelbergh and Reijers (2020) propose in a clearer way a "narrative turn": "We aim to use narrative theory as a theory of technological mediation that conceptualizes how our understanding and actions are mediated by external things (texts, but also other technological objects)" (44). Drawing on Ricoeur's hermeneutics, these authors do not reject the empirical turn and the attention to the materiality of technologies; they seek to understand how this materiality is able to mediate between human beings and the world. The reference to narrative proposes this mediation, inserting technical artifacts into a wider context, that of practices, and therefore connecting technical systems to language, time, and society, that is, collective intentionality. This does not mean, however, that the narrative is separated from such devices. The devices embed narratives; technology is itself a narrative, able to configure time in a story and to reshape human experience. Therefore, the key question is this: How do humans and technologies co-author narrative structures that are able to transform our understanding of our technologically mediated social reality?

According to the "narrative turn," technology has inherent narrative qualities. Technology is narrative in itself and not just because other forms of narrative concerning it are applied to it from the outside. However, this thesis seems paradoxical: What kind of narrative can be intrinsic to a video camera or a computer? At first glance, technology appears to be the opposite of a narrative and its experience of time. Coeckelbergh and Reijers (2020) argue, however, that there are actually deep connections between technology and narrative. This thesis is much more radical than that of Kaplan (2006), according to which narrative theory can be used in order to interpret the way in which humans "read" technology.

Ricoeur himself provides us a clear indication in this regard. Ricoeur (1994) applies the scheme of the three mimeses that we have previously analyzed to the study of architecture. His thesis is very clear: While in literature and historiography the three mimeses apply to the human experience of time, in architecture they apply to space and the act of living.

Let us redefine the three mimeses from an architectural point of view.

- The architectural prefiguration concerns human familiarity with the inhabited space; we build houses because we need first of all a place to live, to "make our space," to install ourselves. Actions are always connected to a space, a *topos*, which gives actions a structure and at the same time is structured by actions.
- The architectural configuration is based on this prefiguration (the need to live) and transforms it. The architectural configuration, which is the project, is a synthesis of the heterogeneous, just like that at work in a literary tale. The only difference is that the unity sought is not temporal but spatial, and therefore it obeys a logic that is not sequential but simultaneous. The architectural work, explains Ricoeur, is a "polyphonic message" that lends itself to a reading that is both "encompassing and analytical" (Ricoeur 1998, 43; my translation). The elements of the synthesis are the multiple variables relating to the act of living (for example, the needs of the population, the history of the site, etc.).
- The architectural reconfiguration is the transformation of the act of inhabiting produced by the configuration. Inhabiting is a reaction to building, that is, to configuration. The inhabitant has a number of expectations that will more or less be met by the project. Thus, just as narrative reconfiguration transforms the reader's temporal experience, the architectural reconfiguration transforms the inhabitant's spatial experience.

Technology in general represents a more fundamental form of narrative than the literary and architectural narratives, which both presuppose it. The technological narrative concerns human agency, that is, the human ability to act in the world. This is equivalent to saying that *any form of emplotment is technology* and vice versa. The limits of technology are the limits of our capacity for emplotment. This statement is in line with the basic principle of Ricoeurian hermeneutics, which sees in the text a model of meaningful action. Ricoeur states that the object of the human sciences is meaningful action and that meaningful action can be understood only by considering the discourse that shapes it, which in turn is fixed in writing and texts: the meaningful action needs to be translated into texts and emplotments. Thanks to the analogy with architecture, we can broaden this thesis by stating that meaningful actions *need to also be translated into artifacts.*

The proponents of the "narrative turn" in post-phenomenology develop this line of research based on an analogical extension of the notion of emplotment. As I said above, the "narrative turn" links the notion of "technical practice" to narrative theory. In doing this, the proponents of the "narrative turn" in post-phenomenology mainly refer to MacIntyre's theory of practice and its reinterpretation in Ricoeur's work *Oneself as Another* (Ricoeur 1992), especially in the so-called little ethics that comprise the last part of the book. Ricoeur's theory of practice is founded on three key notions: action, practice, and life plan. A practice designates an activity that stretches across time. It is a chain of actions which is meaningful only if it is connected to a "life plan," a movement that gathers practices within the wholeness of a human life. Life plans are "vast practical units that make up professional life, family life, leisure time, and so forth" (Ricoeur 1992, 157). The scope of a life plan is to connect the practices to the "unity of a life" (Ricoeur 1992, 160), which is not a coherent and complete whole. Rather, for several reasons, it is a mobile, analog unit. The first is because the narrative of a life never has a single author or a single protagonist; especially at the beginning of our lives, we do not determine the essential changes and choices. Second, a life is never complete because the moments of birth and death are inaccessible by definition and therefore can neither be told nor explained. Third, a life can consist of several narratives, each with its own coherence; narratives conflict with each other, complement each other, and so on. For these reasons, Ricoeur claims that the narrative unity of life cannot be understood as a coherent and complete whole but rather must be read as an "unstable mixture of fabulation and actual experience" (Ricoeur 1992, 162). In *Oneself as Another*, it is the reference to ethics and the idea of a "good life" that provides the key to solving the problem of the coherence and the "achievement" of the narrative unity of life.

Currently the proponents of the "narrative turn" affirm that technology is a set of technical practices that (a) receive their meaning from the narrative unity of life to which they are connected and (b) mediate our practices just like any other type of narrative. Just as humans tell their stories through technology, technology also tells these stories to human beings: "We argue that humans do not only read technologies, but technologies on the other hand "read" humans, insofar as what is experienced by a user must first be constructed in the technology" (Coeckelbergh and Reijers 2020, 87). In terms of the "narrative turn," this means that the narrative function of the technology is a joint operation of the user and the device. The first two phases of *mimesis* correspond to the human activities that use technology as a form of mediation between practices and the narrative unity of life, while the third phase is a joint action of technology and humans. For instance, before a person has ever driven a car, he/she will already have an understanding of the way traffic functions and of the car as a cultural artifact. It is through interaction with the actual car, however, that this prefigured understanding is configured: the understanding of both traffic and the car as an artifact is altered and integrated into a new understanding of the social world (Coeckelbergh and Reijers 2020, 89).

For defending the claim that technologies configure our narrative understanding, the example of software is particularly suitable. Through the instructions it gives to the machine, software creates a narrative, a set of actions that develop over time and have a purpose. The experience made by software (created by the programmer and lived by the user) is a configuring act (mimesis phase 2) because it unites and orders a set of heterogeneous processes in order to adapt to the reader's experience. I would say that *software is the configuring act of any type of narrative embedded in digital technology*. The modeling of an algorithm has a narrative form. In very simple terms, an algorithm is a set of processes that has a beginning, a middle, and an end. One might say that the algorithm is a score that must be played with four hands, by the computer and the user at the same time. A good example is that of the operating system (OS), which is an interface between a computer user and the computer's hardware. An OS is a type of software that performs all a computer's basic tasks, such as file management, memory management, process management, input and output handling, and the control of peripheral devices, including disk drives and printers. It gathers and unifies a series of processes in order to mediate the relationship between users and hardware. Common desktop operating systems include Windows, OS X, and Linux. While each OS is different, most provide a graphical user interface, or GUI, that includes icons and the ability to manage files and folders. They also allow users to install and run programs written for them. Windows and Linux can be installed on standard PC hardware, while OS X is designed to run on Apple systems. The hardware you choose affects what OS(s) you can run.

An OS is therefore an environment in which stories can be developed; the environment confers continuity and purpose to the activities that take place in it. In particular, the OS mediates between the human (social and personal) experience of time and computer time, that is, the computational time, which are completely different. The code, in this case, has a fully narrative function. I am not saying that the code creates a story, like in a video game, but that by mediating between human time and computational time, the OS provides a general configuring act that can be integrated by and connected to a human pre-understanding of time and can thus produce a specific reconfiguration of time. *Software provides a type of narrative schematism, a method for creating narratives in a digital environment.* Computational time, made up of thousands of nested microprocesses, is integrated into human time; the synthesis produced is a mediation that influences and modifies the two types of time. For example, the timing of social networks, the way of living through them, and their speed of communication is a result of this mediation. In producing this synthesis, software makes the computer a universal *narrative* machine. This is perfectly in line with the "narrative turn": "We argue that temporality is configured in the process of technological emplotment in two ways: (1) by means of enforcing or relaxing strict successions of events and (2) by means of 'connectors', which can be dating mechanisms, version control mechanisms, and tracing mechanisms" (Coeckelbergh and Reijers 2020, 98).

## Conclusions

The general thesis of this chapter is that philosophical hermeneutics (especially Ricoeur's work) can be a very powerful tool for understanding what software is. A hermeneutic understanding of software allows us to explain those aspects of software that escape a strictly technical definition, such as the relationship with the user, with the human being, and the social and cultural transformations that software produces, precisely as any text does. The hermeneutic understanding of software considers software itself as a social and cultural product. Furthermore, the hermeneutic circle is a particularly suitable model for describing a dynamic process such as software.

I am also convinced that a hermeneutic understanding of software allows us to clarify the tension between art and science always present in the history of programming. The Ricoeurian dialectic of distanciation and appropriation is perfectly suited to the nature of software, which in fact is explanation (logic, engineering) and understanding (imagination and interaction with the user and modification of agency) at the same time. These two moments are deeply connected: there is no creative imagination without logic and engineering.

## Notes

1. "One common form is computer interface schemes that pose direct questions to the user, such as the ATM machine that displays questions on its screen ('Would you like to make a withdrawal?'), or the 'dialog box' that opens on a computer screen to provide program installation instructions. This is not to claim that we mistake these devices for actual people, but simply that the interface modes take an analogous form" (Rosenberg and Verbeek 2017, 18).

2. Intentionality can be described as the "aboutness" or "directedness" of our conscious states. It indicates the elementary fact that every psychic act has an object, although not necessarily something existent. The phenomenological approach, for Husserl, broadly means the intentional approach. Husserl inherits the concept of intentionality from Franz Brentano, who in turn credits the Scholastics. Husserl begins by specifying what he means by "consciousness," bracketing discussion of the relation of conscious acts to an ego, and focusing exclusively on the intentional character of conscious experiences deriving from Brentano's rediscovery of intentionality (Moran and Cohen 2012, 168–169).
3. "Let us therefore call *low-level*, those languages whose abstract machines are very close to, or coincide with, the physical machine. Starting at the end of the 1940s, these languages were used to program the first computers, but they turned out to be extremely awkward to use. Because the instructions in these languages had to take into account the physical characteristics of the machine, matters that were completely irrelevant to the algorithm had to be considered while writing programs, or in coding algorithms. It must be remembered that often when we speak generically about 'machine language', we mean the language (a low-level one) of a physical machine. A particular low-level language for a physical machine is its *assembly language*, which is a symbolic version of the physical machine (that is, which uses symbols such as ADD, MUL, etc., instead of their associated hardware binary codes). *Programs in assembly language are translated into machine code using a program called an assembler*" (Gabbrielli and Martini 2010, 5).

## References

Achterhuis, H. 2001. *American Philosophy and Technology; The Empirical Turn.* (Indiana Series in the Philosophy of Technology). Indiana University Press.

Amoore, L. 2009. Algorithmic War: Everyday Geographies of the War on Terror. *Antipode* 41: 49–69.

Aristotle. 1902. *Poetics (S. H. Butcher, Ed.). (The Poetic).* London: Macmillan and Co.

Baldwin, R.E. 2016. *The Great Convergence: Information Technology and the New Globalization.* Cambridge, MA: Harvard University Press.

Berry, D. 2011. *The Philosophy of Software.* New York: Palgrave Macmillan.

Carpo, M. 2017. *The Second Digital Turn: Design Beyond Intelligence*. Cambridge, MA: MIT Press.
Ceruzzi, P. 2003. *A History of Modern Computing*. Cambridge, MA: MIT Press.
Chun, W. 2013. *Programmed Visions: Software and Memory*. Cambridge, MA: MIT Press.
Coeckelbergh, M., and W. Reijers. 2020. *Narrative and Technology Ethics*. New York; London: Palgrave Macmillan.
Coeckelbergh, M. 2017. *Using Words and Things: Language and Philosophy of Technology*. New York: Routledge.
———. 2020. *Introduction to Philosophy of Technology*. Oxford University Press.
Colburn, Timothy R. 1999. Software, Abstraction, and Ontology. *The Monist* 82: 3–19.
Elahi, A. 2018. *Computer Systems: Digital Design, Fundamentals of Computer Architecture and Assembly Language*. Berlin: Springer.
Elliott, A. 2018. *AI Culture: Everyday Life and the Digital Revolution*. London; New York: Routledge.
Foucault, M. 1982. *The Archeology of Knowledge*. New York: Vintage.
Frabetti, F. 2015. *Software Theory*. London; New York: Rowman & Littlefield (Media Philosophy).
Fresco, N., and G. Primiero. 2013. Miscomputation. *Philosophy and Technology* 26 (3): 253–272.
Gabbrielli, M., and S. Martini. 2010. *Programming Languages: Principles and Paradigms*. Berlin: Springer.
Golumbia, D. 2009. *The Cultural Logic of Computation*. Cambridge, MA: Harvard University Press.
Hayles, C. 2005. *My Mother Was a Computer: Digital Subjects and Literary Text*. University of Chicago Press.
Ihde, D. 1990. *Technology and Lifeworld: From Garden to Earth*. Bloomington: Indiana University Press.
———. 1999. *Expanding Hermeneutics: Visualism in Science*. Evanston, IL: Northwestern University Press.
Kaplan, D.M. 2006. Paul Ricoeur and the Philosophy of Technology. *Journal of French and Francophone Philosophy* 16 (1/2): 42–56.
Kearney, R. 2006. Ricoeur's Philosophy of Translation. In *On Translation*, ed. P. Ricoeur, 1–24. London; New York: Routledge.
Kittler, F. 2014. *The Truth of the Technological World*. Stanford University Press.
MacIntyre, A. 2007. *After Virtue: A Study in Moral Theory*. 3rd ed. Notre Dame, IN: University of Notre Dame Press.
Manovich, L. 2013. *Software Takes Command*. London: Bloomsbury.

Marino, M. 2020. *Critical Code Studies*. Cambridge, MA: MIT Press.
Mitcham, C. 1994. *Thinking Through Technology: The Path Between Engineering and Philosophy*. University of Chicago Press.
Montfort, N. 2016. *Explanatory Programming for the Arts and Humanities*. Cambridge, MA: MIT Press.
Moran, D., and J. Cohen. 2012. *The Husserl Dictionary*. New York: Continuum.
Pearl, J., and D. Mackenzie. 2018. *The Book of Why: The New Science of Cause and Effect*. New York: Random.
Primiero, G. 2020. *On the Foundations of Computing*. Oxford: Oxford University Press.
Reijers, W., Coeckelbergh, M. 2020. *Narrative and Technology Ethics*. London: Palgrave Macmillan.
Ricoeur, P. 1959. Le symbole donne à penser. *Esprit* 59: 60–76.
———. 1965. *De l'interprétation: Essai sur Freud*. Paris: Seuil.
———. 1969. *Le conflit des interpretations*. Paris: Seuil.
———. 1978. Imagination in Discourse and in Action. *Analecta Husserliana* 7: 3–22.
———. 1983–85. *Temps et récit. 1–3*. Paris: Seuil.
———. 1984. *Time and Narrative 1*. University of Chicago Press.
———. 1986. *Du texte à l'action*. Paris: Seuil.
———. 1992. *Oneself as Another*. University of Chicago Press.
———. 1998. Architecture et narrativité. *Urbanisme* 303: 44–51.
Rosenberg, R., and P.-P. Verbeek. 2017. *Postphenomenological Investigations*. London: Routledge.
Sack, W. 2019. *The Software Arts*. Cambridge, MA: MIT Press.
Thrift, N. 2005. *Knowing Capitalism*. New York: Sage Publications.
Turkle, S. 1984. *The Second Self: Computer and the Human Spirit*. Cambridge, MA: MIT Press.
Turner, R. 2018. *Computational Artifacts: Towards a Philosophy of Computer Science*. Berlin: Springer.
Verbeek, P-P. 2005. *What Things Do Philosophical Reflections on Technology, Agency, and Design*. Penn State University Press.
Verbeek, P-P. 2008. Obstetric ultrasound and the technological mediation of morality: A postphenomenological analysis. *Human Studies* 31: 11–26.
Verbeek, P.-P. 2011. *Moralizing Technology: Understanding and Designing the Morality of Things*. University of Chicago Press.
Wolfram, S. 2002. *A New Kind of Science*. Champaign, IL: Wolfram Research.
Zwier, J., Blok, V. & Lemmens, P. 2016. Phenomenology and the Empirical Turn: a Phenomenological Analysis of Postphenomenology. *Philos. Technol.* 29: 313–333.

# 4

# Software and Writing: A Philosophical and Historical Analysis

## Introduction

In previous chapters of this book I have said that the digital object is a written object—writing is its essence. But why? The digital object (the icons on the screen of my laptop, the apps on my smartphone, the cloud I use to store my documents, the newspaper I read on a tablet, the music I listen to on my phone, or my Instagram account) is a written object, that is, its ultimate essence is writing—it is a string of code. What does this thesis really mean? How to justify it? Why is writing so important? What is the difference between a digital object and any literary text—since both are written objects? In this chapter, I expose one of the central theses of this book: the digital revolution is a revolution of writing, and this revolution has its roots in a long and complex historical process, that is, the history of mathematical writing. A mathematical text is essentially different from a literary text. Symbols work differently. The structure and purpose are different. Writing itself is different.

My thesis is that in the passage from the documentation to the source code to the assembly and finally to the machine code (the levels that I have distinguished in the previous chapter), a "regression to the writing"


L. M. Possati, *Software as Hermeneutics*,
https://doi.org/10.1007/978-3-030-63610-4_4

takes place. This regression to the materiality of writing allows software to work. In fact, it is necessary to consider an essential fact: the machine does not read code like a human being. The human being "sees" a meaning in the code, the machine does not. For the machine there are only traces; that is, electronic signals. For the machine, there is only the writing and re-writing of these signals in its memory. If we consider writing as a set of physical traces corresponding to linguistic structures, then writing is what mediates between the machines and humans and thus allows software to function. In the passage from one level to another, software loses its meaning and remains pure writing.

Another point has to be considered. Writing is not a passive, neutral object. It implies in itself a hermeneutic process, a visual and graphic intelligence, that makes a more refined and complex treatment of information possible. The hermeneutics of the text we talked about in the previous chapter is based on this hermeneutics of writing. The hermeneutic process of appropriation and distancing is rooted in a much deeper hermeneutic level, which I define as a "material hermeneutic" (the expression was coined by Ihde 1990) because it concerns writing as a graphic object. Writing is not a set of "dead" signs but a "living matter," which has its own autonomy and is capable of making us discover new things in the world. This material hermeneutics is almost lost in the literary text, while it reaches its maximum expression in mathematical language and software.

I justify these theses in three ways: (a) a preliminary analysis of Ong's theses on writing and computation; (b) a historical analysis of the birth of the modern mathematical language between the seventeenth and eighteenth centuries; (c) a historical analysis of the development of software. The "hero" of this chapter is Descartes, inventor of modern mathematical writing.

## Orality and Literacy

Writing opened a completely new phase in the history of human thought. Without writing, the human mind could not fully exploit its potential. This is the central thesis of *Orality and Literacy* by Ong (1982), a fundamental text for the study of writing, the philosophical meaning of writing, and in general the theory of media. Accustomed to an essentially written culture, in which printing is widespread, we cannot conceive of a world without these media and a way of thinking entirely based on orality—what Ong calls "primary orality." Influenced by McLuhan, Parry, and Havelock, Ong claims that the constitution of the human mind is based on language and communication. Therefore, understanding how communication processes take place is essential to understanding the human mind. Wherever there are humans, there are language and communication, which are the only sources of meaning. However, language is not something abstract: it is embodied, materialized, and realized through technologies. These technologies shape human consciousness and influence our way of understanding truth and meaning. They are not neutral at all.

Ong speaks of primary orality to indicate a situation totally devoid of writing. Primary orality is a form of linguistic communication based on the immediate physical presence of sound, and so it does not know any form of recording. "Sound is always 'alive' indicating that some physical action is going on" (Ong 2002, 541). The preservation of the meaning depends on the mnemonic effort. As the Homeric poems show, a primary orality culture is circular and redundant because it is based on fixed mnemonic schemes such as formulas, epithets, or the structure of the verse, which are a fixed way of processing and transmitting experience data and information. In primary orality there is no reference to any syntactic structure and creativity; everything is based on iteration. For this reason, primary orality is more aggregative than analytical, as Ong holds:

> The elements of orally based thought and expression tend to be not so much simple integers as clusters of integers, such as parallel terms or phrases or clauses, antithetical terms or phrases or clauses, epithets. Oral folk prefer, especially in formal discourse, not the soldier, but the brave soldier; not the princess, but the beautiful princess; not the oak, but the sturdy oak. Oral expression thus carries a load of epithets and other formulary baggage which high literacy rejects as cumbersome and tiresomely redundant because of its aggregative weight. (Ong 1982, 38)

Primary orality is therefore conservative by nature; it tends to strengthen and perpetuate the given order, the habitus of a given society. It is less inclined to experimentation and abstraction. An oral culture also gives more importance to competition, both in language and in lifestyle. A characteristic expression of primary orality is in fact rhetoric, and so the tendency is toward emphasis, public participation, the ability to kindle the emotions of the people who listen, and convey consent and a sense of community as well. "When a speaker is addressing an audience, the members of the audience normally become a unity, with themselves and with the speaker" (Ong 1982, 72). In other words, primary orality is a form of thought connected to the present and the given situation, to the here and now—it confirms a situation, the *status quo*. For this reason,

> the oral mind is uninterested in definitions. Words acquire their meanings only from their always insistent actual habitat, which is not, as in a dictionary, simply other words, but includes also gestures, vocal inflections, facial expression, and the entire human, existential setting in which the real, spoken word always occurs. Word meanings come continuously out of the present, though past meanings of course have shaped the present meaning in many and varied ways, no longer recognized. (Ong 1982, 46)

In an oral culture, words acquire meaning only from the immediate situation, the performance. There are no dictionaries or fixed definitions of the terms. As Ong claims, having no definitions means being unable to distinguish things from each other, and therefore being unable to distinguish oneself from the rest of the world. Primary orality, therefore, lacks self-analysis; it is unable to process high levels of complexity with adequate accuracy. This attitude responds to the unifying and centralized economy that is characteristic of sound, which is the true phenomenological trait of primary orality.

> A sound-dominated verbal economy is consonant with aggregative (harmonizing) tendencies rather than with analytic, dissecting tendencies (which would come with the inscribed, visualized word: vision is a dissecting sense). It is consonant also with the conservative holism (the homeostatic present that must be kept intact, the formulary expressions that must

> be kept intact), with situational thinking (again holistic, with human action at the center) rather than abstract thinking, with a certain humanistic organization of knowledge around the actions of human and anthropomorphic beings, interiorized persons, rather than around impersonal things. (Ong 1982, 71–72)

Ong's thesis is that, even though it is based on primary orality, writing has radically transformed the structure of oral culture, thus opening a new phase in the history of human thought. Without writing, the literate mind could not think as it does, "not only when engaged in writing but normally even when it is composing its thoughts in oral form. More than any other single invention, writing has transformed human consciousness" (Ong 1982, 77). Hence, we are dealing with a perfectly symmetrical situation: the writing has the same characteristics as primary orality, but reversed. I do not want to analyze them one by one. I will just summarize them in Table 4.1 by following Ong's suggestions:

The contrast between orality and writing and the transformation that writing has imposed on oral culture are evident in Plato. Plato's thesis is that writing is inhuman because it pretends to recreate outside the mind what can actually exist only inside it. Writing is a passive and inanimate thing; there is no intention or interaction in it. In writing, the essence of thought is lost; memory weakens because it depends on external objects. Nonetheless, Plato writes. Indeed, his *Dialogues* are an unsurpassed model of writing. As Havelock (1963) demonstrated, despite the criticism of

**Table 4.1** The characteristics of writing and orality compared

| Orality | Writing |
|---|---|
| Modularity | Continuity |
| Parataxis | Hypotaxis |
| Aggregative | Analytic |
| Redundant | Economic |
| Conservative | Experimental |
| Close to concrete human life | Distant from concrete human life |
| Homeostatic | Changing |
| Agonistic | Non-agonistic |
| Social reading | Solitary reading |
| Sound: inner reality | Vision: external reality |

writing in the *Phaedrus*, all Platonic epistemology is based on the rejection of the old oral culture represented by Homeric poems. Plato does not admit poets in *The Republic* because he thinks that ideas are true knowledge. But ideas are essentially immobile and abstract visual objects. They are the opposite of orality.

> The term idea, form, is visually based, coming from the same root as the Latin video, to see, and such English derivatives as vision, visible, or videotape. Platonic form was form conceived of by analogy with visible form. The Platonic ideas are voiceless, immobile, devoid of all warmth, not interactive but isolated, not part of the human lifeworld at all but utterly above and beyond it. Plato of course was not at all fully aware of the unconscious forces at work in his psyche to produce this reaction, or overreaction, of the literate person to lingering, retardant orality. (Ong 1982, 79)

Plato was the first true thinker of writing, the one who best expressed the cognitive potential of writing. The idea is a written object because it would be inconceivable without writing, without the objectivity, analytical power, and abstraction of writing. Writing allows the human mind to reach the level of abstraction necessary to think of ideas. Moreover, it allows the mind to store information over time—therefore, it makes bureaucracy and history possible.

How can writing restructure thought starting from primary orality? First of all, Ong claims that writing differs from orality because it is a "technologicalization" of language. Writing is a technology that translates sound into a visual space. It transforms our perception of language through the use of specific artifacts—papyrus, parchments, pens, inks, and so on. "A text is a technological product, a physical object distinct from the person producing it" (Ong 2002, 542).

By creating an abstract space, writing fragments and reconstructs the human experience. It is an architectural intelligence that is based on the exploitation of the semantic and cognitive resources of space (Knauff 2013). It allows the first form of digitization or discretization of the experience: "Alphabetic systems are forms of digitization" (Ong 2002, 544). The use of space and vision allows the gathering of more information, that is, a greater number of elements and connections between these elements. A visual representation admits not only a sequential order, like sound, but also a simultaneous, parallel order. Furthermore, in writing, the use of space is significant, that is, it contributes to the construction of meaning—think, for example, of the page, the distance of the words between them, the size of the words, the color, and the possibility of integrating a text with diagrams and pictures. In printing, the layout has a semantic dimension.

Ong's thesis is also confirmed by Wolf (2007, 2018) according to which writing and reading have made possible the development of new neurological structures in the human brain. The reading brain is not the listening brain. It is a brain capable of processing and analyzing a greater quantity of very complex data. Wolf shows that reading is not a natural attitude of humans, but an invention that dates back to 6000 years ago in Mesopotamia, with the cuneiform writing of the Sumerians. In order to learn to read, the human brain had to create sophisticated connections between neuronal structures and circuits originally responsible for other more basic processes, such as sight and spoken language. The brain is reshaped in a new way by reading. At both the biological and intellectual levels, reading allows the human species to go beyond the information already given in order to produce countless and wonderful new thoughts. Culture as we know it now is the daughter of the brain that writes and reads.

In the evolution of writing, the alphabet and printing have represented two crucial revolutions, which have had strong cognitive consequences. As McLuhan (1962) has shown, movable type printing has enormously changed the structure of human consciousness. With the press, the word is now considered an object reproducible in an exact way. The text itself becomes exactly reproducible. It, therefore, acquires an identity in a strong sense, a closure, the impression of being something finite, complete, a unity in itself. Furthermore, with the passage from handwriting to printing the visual character of writing is strengthened:

> […] Writing moves words from the sound world to a world of visual space but print locks words into position in this space. […] 'Composing' type by hand (the original form of typesetting) consists in positioning by hand preformed letter types, which, after use, are carefully repositioned, redis-

> tributed for future use into their proper compartments in the case (capitals or 'upper case' letters in the upper compartments, small or 'lower case' letters in the lower compartments). Composing on the linotype consists in using a machine to position the separate matrices for individual lines so that a line of type can be cast from the properly positioned matrices. Composing on a computer terminal or word processer positions electronic patterns (letters) previously programmed into the computer. (Ong 1982, 119)

Ong connects the exact reproducibility of texts made possible by movable type printing to the scientific revolution. The printed texts in fact allow the connection between the exact observation of the phenomena and the exact verbalization of this experience. Technical prints and technical verbalization reinforced and improved each other. The age of writing and printing has produced a new form of orality, the "secondary orality," that is, "the orality produced by machines devised through the use of reading and writing and print—radio, television, and associated mechanical creations" (Ong 2002, 521).

It is difficult to overestimate the importance and influence of Ong for studies on writing and on the relationship between orality and writing. I wanted to thoroughly analyze his arguments because it helps me to highlight a crucial aspect of my research: the connection between human consciousness and technology. However, I also want to advance two criticisms of Ong's approach to the study of writing. I will analyze these criticisms (and answer to them) in the next sections of this chapter:

1. Ong does not grasp all the philosophical consequences of connecting human consciousness and technology. "There is nothing more natural for a human being than to be artificial" (Ong 2002, 541). Writing is an artifact, a technology, and requires specific tools. Ong states that literature and logic are a result of the evolution of the mind that writes and reads. The scope of this thesis is enormous: there is no clear distinction between the subject and the object, between culture and nature, but a continuous process of translation and mediation, and technology—that is, the production of artifacts, of hybrids that they are neither nature nor culture—is at the center of this mediation. We

cannot fully understand the essence of writing if we do not first understand the conceptual couple subject/object from another perspective. I will try to fill this lack in Ong's approach in the next section by drawing on Latour's anthropology of science.

2. Ong only considers writing systems connected to the alphabet, therefore based on the imitation of oral language. On the other hand, he does not consider "pure" writing systems, which were not built to translate sounds into spatial relationships. In reality, the question is very complex: Ong talks about it in a paper (see Ong 2002, 54–76), where he shows that the evolution of mathematical writing systems has been long and complex, and not completely separated from primary orality. However, it seems to me that Ong does not grasp that mathematical writing is substantially different from verbal or alphabetic writing. Mathematical writing is not the translation of sounds in an abstract space, but the expression of completely abstract ideas, numbers, and figures. Mathematical writing, therefore, concerns objects that are completely different from those designated by verbal or alphabetic writing. Not only that: the purpose of mathematical writing is not simply to give verbal form to abstract ideas, but to operate with numbers. Writing for the mathematician is not a means of expression and communication; it is also this, but only secondarily. For the mathematician, writing is above all a work tool that has a heuristic power, that is, it is capable of discovering new mathematical entities. In mathematics, writing is not just a means of communication, but an active part of the construction of objects and truths. I will try to fill this lack in Ong's approach in the fourth and fifth sections of this chapter, which have a more historical character.

## The Revolution of Printing

In contrast to Ong's research on the shift from an oral to a written culture, Eisenstein (1983) focuses on the historical transition from handwriting to printing among the educated elites of Western Europe—in particular, the ways that print altered written communication within the republic of letters. Unlike McLuhan (1962), Eisenstein emphasizes the complexity of this historical change and the difficulty of translating it into a single theoretical formula. For this reason, while acknowledging the value of McLuhan's research, Eisenstein focuses on how the introduction of printing influenced different social groups, particularly the elites, in Western Europe between 1400 and 1500.

Eisenstein does not analyze the transition from an oral popular culture to a printed culture or the spread of literacy due to writing; instead, she scrutinizes a revolution within the revolution. The press is a factor—a decisive factor but not the only one—in the profound transformation of European culture between the fifteenth and sixteenth centuries. The merit of Eisenstein's research is that it not only historicizes the theoretical schemes of Ong and McLuhan, but, above all, it illuminates a crucial phase in the history of writing by going beyond overly static schemes (e.g., the dualism of orality–writing) and exclusive focuses (e.g., the theme of communication). In the Western world, writing has a very particular history, and the advent of printing represents a fundamental turning point. As I later try to show, the advent of software is another milestone in the long history of writing that has radically transformed typographic culture. I venture that the advent of software—and, therefore, the transformation of computers from simple mathematical machines into universal machines capable of adapting to any human task—represents a mental revolution that is comparable to that of typography and has effects that are still not completely clear. A profound connection certainly exists between the advent of software and the cultural change between the end of the twentieth and the beginning of the twenty-first century, but this relationship remains obscure and has not been fully comprehended. Thoroughly understanding this problem is one of the central conditions of "critical code thinking," which I will discuss in the conclusions of this book.

Precisely defining the advent of printing is a highly complex undertaking; "we must single it out as an event which was sui generis and to which conventional models of historical change cannot be applied" (Eisenstein 1983, 128). An indisputable historical fact, though, does exist: over a few decades beginning in the second half of the 1400s, printing machineries and laboratories were built in major urban centers throughout Europe and produced books in almost all the languages of Western Europe.

Important figures such as Peter Shoeffer, Vespasiano da Bisticci, and Aldo Manunzio created new work and a new form of culture. The printing laboratories of the late 1400s were modern types of businesses and centers of culture in all respects. If the scriptorium was a place of meditation and prayer, the printing laboratory was the site of commerce, money, and even workers' agitation against the bosses. Books, even the consecrated ones, were printed not only for love of God but for the purpose of being sold every day. The social and psychological consequences of this change clearly appeared as early as 1500.

As mentioned, attempts to summarize the changes produced by the press in a single formula are misleading. The historical point of view requires overcoming sterile dualisms such as the view that sees the passage from the handwritten word to the printed word as the transition from a "culture of the image" to a "culture of the word" (Eisenstein 1983, 45). By making it possible to dispense with images for mnemonic purposes, the press reinforced the iconoclastic tendencies already present in many Christians; Calvin provides a perfect example of the defense of reading in contrast to images (Eisenstein 1983, 45). However, the press also contributed to the transformation of images and their meanings. Significant examples of this shift are Dürer, Cranach, and Holbein, who developed new printing techniques for images. In some fields of knowledge, such as architecture, geometry, and geography, printing changed and increased the functions of images. The possibility of reproducing identical images was fundamental to the development of science and pedagogical and musical books. Nonetheless, we should not think that the spoken and the handwritten word were simply reduced to silence: the fight between the cultural and social models of scribes and booksellers remained acute for at least two centuries. The press transformed the earlier amanuensis culture and strongly influenced the Renaissance. One of Eisenstein's crucial theses is that the press played a key role in the constitution of the new sense of the past that characterized the Renaissance and the passage to the modern age. Through the production of maps, chronologies, catalogs, and new editions, printing created for scholars a space-time reference structure that was less rhetorical, more shared, and more rigorous than in previous eras.

I want to indicate particular characteristics of the press that seem important to analyzing and understanding the meaning of this revolution. The first effects of printing between 1400 and 1500 were to increase the production of texts and modify reading practices. The era of the glossator and the commentator ended. New scholars had many different books at their disposal and, instead of focusing on only one, could decide to compare references from among them. In the sixteenth century, access to a greater quantity of cheap books created a new cultural ferment and produced a completely different vision of culture than in previous eras. An amanuensis had few books at his disposal and usually a very compact, unitary, monolithic conception of knowledge and history. In contrast, the new intellectual of the seventeenth century had a much more critical, multifaceted vision of knowledge, marked by conflicts and diversity. Fuller bookshelves increased the possibilities to consult and compare different texts. As Eisenstein points out, giving the example of Montaigne:

> It would be foolish to assert that the most newsworthy events of the age made no impression on so sensitive an observer as Montaigne. But it also seems misguided to overlook the event that impinged most directly on his favorite observation post. That he could see more books by spending a few months in his tower study than earlier scholars had seen after a lifetime of travel also needs to be taken into account. In explaining why Montaigne perceived greater "conflict and diversity" in the works he consulted than had medieval commentators in an earlier age, something should be said about the increased number of texts he had at hand. (Eisenstein 1983, 48)

Greater production and availability of texts also meant greater possibilities for combining different knowledge and theories and undertaking cultural exchanges, or intertwining different social worlds and traditions. As Eisenstein writes, "once old texts came together within the same study, diverse systems of ideas and special disciplines could be combined;" in fact, "increased output directed at relatively stable markets, in short, created conditions that favored new combinations of old ideas at first and then, later, the creation of entirely new systems of thought" (49). For example, printing created a new profession, the printer, and a new professional community of those working in printing houses who acquired different skills: type founders, proofreaders, editors, illustrators, translators, index extenders, and traders. "The decisions made by early printers, however, directly affected both tool making and symbol making. Their products reshaped powers to manipulate objects, to perceive and think about varied phenomena" (Eisenstein 1983, 70). On a strictly cultural level, the first century of printing created the basis for broad, precise knowledge of the sources of Western, classical, and Christian thought. In a certain sense, printing an ancient text also required establishing its exact content in a clear, usable way.

The second characteristic of printing is standardization. Printing first made it possible to publish large quantities of almost identical copies of the same text, resulting in the standardization of texts and, more importantly, images. Maps, tables, and diagrams but also dictionaries, grammar, and reference manuals became ordered, systematically organized, rationalized, codified works. Producing two identical copies of an image was much more difficult when working by hand than by printing. The spellings and variations due to individual scribes disappeared. A new type of individual, a sense of the private, and a new model of reading emerged. In fact, one effect of the standardization of writing was to free the writing individual from all the restrictive ideals (in most cases of a devotional nature) that demanded the disappearance of the amateurs "behind" the text. Montaigne, once again, provides an important example: "By presenting himself, in all modesty, as an atypical individual and by portraying with loving care every one of his peculiarities, Montaigne brought this private self out of hiding, so to speak. He displayed it for public inspection in a deliberate way for the first time" (Eisenstein 1983, 62). However, the development of standardization also had an opposite effect; it eased the spread of stereotypes, such as the noble, the citizen, and the peasant.

The third characteristic of printing I want to mention is the reorganization of data and mental models, which Ong and McLuhan stress but not from a historical perspective. According to Eisenstein, typographic standardization required using catalogs and indexes that were always the same and uniform. The work of lexicographers and, therefore, the production of increasingly precise and systematic lexicons should not be forgotten either. Printing amplified and improved practices (e.g., cataloging, consultation, and indexing) already present in the Middle Ages. The printing of works also played a significant role by unifying mutually reinforcing intellectual and commercial activities. The introduction of the press led to the production of catalogs, dictionaries, atlases, and other reference books of enormously better quality than in previous centuries. Entire fields of study were rearranged according to the logic of the printed page, with its spaces, titles, citations, page numbering, regular punctuation, indexes, arrangement of elements, and divisions into paragraphs and chapters. The "graphic reason" that emerged turned out to be immensely powerful. Due to the press, important legal instruments took on more precise, effective forms, and many professions and trades experienced a great phase of innovation. Moreover, printing allowed correcting errors in subsequent editions. Large-scale data collection became subject to new forms of feedback not possible in the era of scribes; cooperation, exchange, and communication between writers, scientists, and publishers increased (Eisenstein 1983, 85).

The fourth noteworthy characteristic of printing is the end of patronage and the birth of a specific social class: intellectuals, particularly authors. The press birthed a new, autonomous social group composed of professionals engaged in producing and distributing printed materials. Those who took advantage of the new careers opened to the talents of skillful writers "were not disembodied spirits who must be materialized to be believed. They were, rather, complex flesh-and-blood human beings. (Some forty-odd volumes were required to cover the lives of the more celebrated inhabitants of the 'Republic of Letters' by the mid-eighteenth century)" (Eisenstein 1983, 111). Moreover, physical foundries, workshops, and offices were built to serve the needs of this "fictitious realm." Profits were made by tapping the talents that gravitated to it. Printing houses served not only as industries but also as multilingual communication centers diffusing an ecumenical, tolerant ethic. They were the most dynamic part of society at the time. These intellectuals had full contact with the deep dynamics of the societies of their time and knew the art of advertising well. By no means were they isolated scholars huddled over their tables.

However, by far, the most peculiar phenomenon introduced by the press concerns the conservation of written material. This phenomenon is the condition of possibility for all the preceding phenomena and is the most complex to reconstruct. As Eisenstein points out, the transformations introduced by the press should not be underestimated, but neither should they be overestimated.

Texts from the era of scribes had very low durability: "No manuscript, however useful as a reference guide, could be preserved for long without undergoing corruption by copyists, and even this sort of 'preservation' rested precariously on the shifting demands of local elites and a fluctuating incidence of trained scribal labor. [...] More than one record required copying, which led to textual drift. Durable records called for durable materials. Stone inscriptions endured; papyrus records crumbled" (Eisenstein 1983, 87). Printing radically changed this logic: conservation was guaranteed by the number of editions—the quantity not the quality of the single manuscript. The printed book often had poor quality and disintegrated much faster than the manuscript, but its copies and diffusion guaranteed its longevity and pervasiveness. Here emerged a fundamental character of typographic culture: accumulation. Standardization, continuity, and cumulativity were closely linked. Printers, at least initially, produced not new books or contents but a greater quantity of existing books—replicating amanuensis culture transformed it.

The increase in copies spurred circulation, democratization, and invention. The stable conservation of acquired knowledge was the prerequisite for creation and change. The tasks of memorizing and copying consumed less energy. Moreover, conservation and typographical stability had highly important geopolitical effects—evident not only in the stability of geographical maps that fixed the borders of the states but also in the radicalization of language barriers, standardization of languages, creation of national literary cultures, and marginalization of minority dialects. These phenomena also had interesting religious consequences: the spread of the press caused the fragmentation of Latin Christianity. As independent lay printing houses reproduced documents concerning the liturgy, canon law, and even theology, escaping the control of the papal curia became easier, and the clergy became much more susceptible to manipulation by political powers able to influence the work of printing houses. The press gave writing an autonomous psychological and political authority. Printed content was perceived as authoritative, enduring, and reliable, bestowing immortality. Consequently, as the role of collective authority and the cult of tradition disappeared, the press created new forms of attribution of texts and affirmed the theme of property rights.

In short, the press constituted a genuine mental, cultural, and social revolution. This point of view raises interest in how Eisenstein reinterprets the Renaissance period, or the revival of classical studies and the development of what is called "a modern sense of individuality." Criticizing traditional analyses (Burckhardt; Panofsky), Eisenstein underlines the centrality of the press in the birth of a "new man" embodied above all by the figure of the master printer, who was simultaneously a scholar and a technician capable of also being an advertiser. In Italy, this "new man" rediscovered the classics, while in Germany, he was the protagonist of Protestant reform. The new powers of the press shaped the Protestant reform from the outset. The reformers realized the usefulness of the printing press to their cause and recognized its importance in their writings.

In short, seven main phenomena introduced by printing can be identified:

- Increased production and popularization of books
- Modified use of books
- Content standardization
- Cumulative data
- Data retention
- Reorganization of data
- Creation of new professionals

## The Symbolic Revolution: Part 1

This thesis (writing as a diagram, therefore as a heuristic power) is confirmed by the history of the symbolic revolution that took place between the seventeenth and eighteenth centuries in mathematics, starting precisely with Viète, Descartes, and Leibniz. Serfati (2005) uses the expression "symbolic revolution" to indicate the process by which writing became the main locus of mathematical invention and discovery, completely distinguishing itself from any other form of language.

After 1637, with the publication of Descartes' *Géométrie*, writing became more and more essential to the work of the mathematician—that is, the familiarity with a writing completely separated from oral language and rhetoric. Viète, Descartes, and Leibniz invented a new form of writing which constructed at the same time (a) its object, hence its content; (b) the operation on the object, that is, its transformation starting from a series of rules; (c) the interpretation of both, that is, their meaning in relation to the rest of mathematics. A fourth was then added to these three aspects: (d) combinatorics, that is, the automatic character of mathematical writing—the material concatenation of symbols precedes and conditions the discovery of the object and its transformation, and this concatenation is autonomous from the will of the subject who uses that script or who reads it. The birth of this new mathematical writing was by no means simple, nor was it the planned creation of a single school or a cohesive group of thinkers. As Serfati (2005, 34–36) recalls, the very protagonists of this turning point were not fully aware of the importance of their invention and the enormous consequences it would have on the development of mathematics. Writing would come to life and exceed the expectations of its inventors, leading to the discovery of new mathematics. There would no be longer any border between invention and discovery, nor between meaning and signifier, nor between language and object. It was a strange, unprecedented phenomenon, which today we find hard to understand because we are too used to the modern mathematical language.

## Before Descartes

Until the seventeenth century, mathematics spoke in natural language. The writings of mathematicians were not filled with formulas, but were written in spoken language: problems, proofs, and procedures were explained by figures and long explanations. Euclid's *Elements* did not feature symbolic writing as we know it today. The text was meant to be read and interpreted like any other text. There were symbols (the letters A, B, C, etc.) but they only served as abbreviations. Euclid's work was kept on a rhetorical, descriptive level. Proof of theorems was always developed in geometric form, that is, through figures, not formulas. Only with Diophantus did the use of a special symbolic develop to address and resolve purely numerical issues with purely numerical methods (Serfati 2005, 57; Cajori 1993, 71–74; to delve into the history of Greek mathematics, Heath 1981 is indispensable; see also Katz 2009). Signs are not only used to designate parts of a figure. They are used to discover a new object and modify it.

The prevalence of natural language is also evident in Cardan, who is a very representative author of sixteenth-century mathematics. In his book *Ars Magna* (1545), Cardan uses natural language and figures to describe problems and solutions. Everything is described clearly and completely, leaving no room for interpretation. Cardan belongs to the group of Italian algebraists of the sixteenth century whose main representative is Luca Pacioli. Although natural language was still dominant, the writing of these mathematicians is full of abbreviations, such as p and m which are used as abbreviations for addition and subtraction; R instead designates the root while "co" the unknown, and "ae" stands for "aequalis," which in Latin means equal. It is therefore still a hybrid writing: neither purely natural nor purely symbolic. Cardan has the merit of introducing the solutions of the equations of the third and fourth degree, of which, however, he was not the discoverer, as he himself admits; he had in fact taken up the idea of Niccolò Tartaglia and Ludovico Ferrari, two other important mathematicians of the time.

In this hybrid writing, the sign still has a passive value and is at the service of natural language. As Cajori (1993, 85) shows, Cardan used two distinct symbols to indicate the operations of addition and subtraction. He took up other symbols for the powers from Pacioli. For example, in his *Ars Magna*, he used the phrase "Rem ignotam, quam vocamus positionem" to designate the unknown, that is, the indeterminate, variable part of the problem.

The case of the sign of equality is very significant. Equality was indicated by a series of expressions such as *aequales, aequantur, esgale, faciunt, ghelijck,* or *gleich*, and sometimes the shortened form *aeq*. This way of writing is found in Keplero, Galileo, Torricelli, Cavalieri, Pascal, Napier, Briggs, Gregory St. Vincent, Tacquet, and Fermat (Cajori 1993, 297). Robert Recorde was the first to use the symbol we know today, =. The use of this symbol to indicate equality was imposed at the end of a long process. The same symbol, for example, was used by François Viète in his *In artem analyticem isagoge* (1591) to indicate the mathematical difference. In 1638, Descartes used it to indicate *plus or minus* (Cajori 1993, 298). The symbol also had to outperform the competition from many others. The main "adversary" was precisely the symbol introduced by Descartes in his *Géométrie*. This symbol "is simply the astronomical symbol for Taurus, placed sideways, with the opening turned to the left. This symbol occurs regularly in astronomical works and was therefore available in some of the printing offices" (Cajori 1993, 301–302; see also Serfati 2005, 239). Descartes' symbol spread widely in Holland and England, albeit often with some variations. The "victory" of the sign = occurred in the eighteenth century. The Recorde symbol was used by John Wallis, Isaac Barrow, and Isaac Newton. It was Leibniz, finally, who established the dominance.

Obviously, this is only one example of a much more complex and stratified history, which we can only hint at here.

## Descartes, and Beyond

As Manders (2006) notes, Descartes learned mathematics from Clavius' texts and was very familiar with the techniques of the German school of mathematics, the so-called cossist algebra. These mathematicians too, like the Italians, had a hybrid rhetorical and symbolic writing. Descartes understood the intrinsic limits—expressive and operational—of this writing and overcame them by inventing a new notation. In this, he was also helped by the work of another mathematician: François Viète.

Now, the real turning point occurred not in the *Géométrie* (Descartes 1954) but in the *Regulae ad directionem ingenii* (*Rules for the Direction of the Mind*), one of Descartes' earliest major writings, included in the posthumous inventory of his effects. The book is about the proper method of knowing and the guidance of the mind in thinking and discovery. It was planned to be in three parts, each consisting of twelve rule-rubrics plus elaboration, but only the first twenty-one rubrics exist (the last three without elaboration). "The whole treatise, according to a general plan exposed in Rule XII, was supposed to contain a description of a general method to solve any question and was intended to consist of three parts of twelve rules each, dealing respectively with 'simple propositions,' 'perfect questions,' and 'imperfect questions'" (Rabouin 2010, 432).

In *Rule* XIV Descartes introduced the foundation of his geometrical calculus by claiming that "in all reasoning it is only by means of comparison that we attain an exact knowledge of the truth" and that "the business of human reason consists almost entirely in preparing this operation." This established the central role of proportion in the treatment of any "question" and the main goal of the method in this second part of the treatise: to reduce these proportions to simple comparisons, that is, equalities (Rabouin 2010, 441). Within this general structure, any "question" can be expressed in terms of proportions between magnitudes and, in consequence, through some spatial representation since no other subject [than spatial extension] displays more distinctly all the various differences in proportions. Therefore, Descartes claimed that "perfectly determinate problems present hardly any difficulty at all, save that of expressing proportions in the form of equalities, and also that everything in which we encounter just this difficulty can easily be, and ought to be, separated from every other subject and then expressed in terms of extension and figures. Accordingly, we shall dismiss everything else from our thoughts and deal exclusively with these until we reach *Rule* twenty-five."

In *Regula* XVI (they are 21 in all) Descartes said of writing:

> Besides, as, among the innumerable dimensions which can imagine ourselves in our imaginations, we said that we cannot kiss more than two at the same time, with one look, either with the eyes or with the mind, it is good to remember all others exactly enough so that they can present themselves to us whenever we need it. It is for this purpose that the nature seems to have given us memory; but like her is often prone to fail, and in order not to be forced to give part of our attention to renewing it, while we are busy with other thoughts, art is very timely invented writing, with the help of which, without giving anything to our memory, and giving up our imagination freely and without sharing with the ideas that occupy him, we entrust to the paper what we will want to retain [...]. (Descartes 2018, 35)

In this passage, Descartes stresses the importance of writing for memory. The thesis is that writing is necessary due to the weakness of human memory. Writing allows clarity and helps the attention to focus on single things and to remember them in memory. This then allows you to retrace all the things you think about in a faster and more agile way. Now, for Descartes, the best form of writing—as he underlined in the rest of the text—must be composed of "very short signs" so that "after having distinctly observed the individual things we can go through them all with a very fast movement of thought and intuit the greater part of them together" (Descartes 2018, 35). Here we already find a crucial idea: writing is not a passive aid to thinking. Instead, it is an active tool, which guides thought and allows it to go where it otherwise cannot. In the next passage Descartes introduces the new symbolic:

> All that should be considered as the unit, for the solution of the question, we will designate it by a single note, that can be taken arbitrarily. But for convenience, we will use the characters a, b, c, etc., to express the quantities already known, and A, B, C, for unknown quantities, which we will precede by digits 1, 2, 3, 4, etc., to indicate the number, and follow the same numbers to express the number of relationships they contain. For example, if I write 2 to 3, it is as if I say, double the size represented by a, which contains three reports. By this means, not only we will save words, but still, what is capital, we present the terms of difficulty so bare and so cleared, that even if we do not forget anything useful, we will not leave however nothing that is superfluous [...]. (Descartes 2018, 36)

In this passage there are two aspects to underline: (a) the use of numbers and letters, and (b) the representation of the power through the exponent. Descartes here took up the innovation introduced by Viète, namely the use of letters to indicate both the indeterminate part and the unknown part (variable) of the equation (Serfati 2005, 157). The indeterminate data is indicated with lowercase letters, while the unknown is indicated with capital letters. The use of letters therefore allows a greater level of abstraction. However, there is another aspect that emerges from this passage clearly: Descartes conceived this new writing in an operational way, that is, as a tool that allows the mathematician to get to know his object better. The symbolic writing must represent (a) the mathematical object, that is, the meaning; (b) the internal and external relations to the concept; (c) the operations to be applied to the object according to some rules, hence the way in which the object is transformed; (d) it also provides an interpretation of a, b, and c, in the sense that it defines them. Writing, therefore, has a descriptive and operational function at the same time; by describing its object, it constructs and transforms it. For Descartes, it was absolutely evident that the knowledge of the mathematical object depends on—and passes through—writing. There is no mathematical object in itself; the meaning of the mathematical text is inseparable from the signifier (the writing) and therefore from combinatorics (Serfati 2005, 183). *The mathematical object results from the interpretation of the self-constituted symbolic form.* This principle would become evident and fundamental for Leibniz.

This awareness emerges not by chance in the *Regulae*. It is in fact in the *Regulae* that Descartes clearly developed for the first time his philosophical project of the *mathesis universalis*, defined as follows in a letter to Beeckmann of 26 March 1619:

> A science with new foundations, making it possible to solve in general all the questions that one can propose in any kind of quantity, as well continuous as discontinuous, but each according to its nature [...] incredibly ambitious program. (Descartes 1988, 38–39)

I do not want to analyze the complex question of the philosophical meaning of the project of the *mathesis universalis*; I limit myself to referring to Rabouin (2010). This is not the subject of this book. As Rabouin claims, the *mathesis universalis* is "a discipline that Descartes studied before he dared to 'tackle the somewhat more advanced sciences,' according to a rule that states that one should never go beyond the easiest matter 'till there seems to be nothing further which is worth achieving where they are concerned'—the exact contrary of a program indeed" (2010, 435). The *Géométrie* cannot therefore be seen as the most complete expression of the new *mathesis universalis*.

The *Géométrie* is a very important text in the history of mathematics (for a presentation of the contents of the work, see Serfati 2005).

Analyzing the place of this work in the history of mathematics of the time is a very complex undertaking (I refer to Israel 1997; Julien 1996). What interests me here is the role this work played in the development of mathematical writing—the impact on the way mathematicians write, that is, the use of symbols. The *Géométrie* represents a sort of Rosetta Stone to which all subsequent mathematicians starting from Leibniz will refer.

The book announced a new approach in mathematical thinking. Descartes boasted in his introduction: "Any problem in geometry can easily be reduced to such terms that a knowledge of the length of certain straight lines is sufficient for construction" (Descartes 1954, 13). He then proceeded to show how arithmetic, algebra, and geometry could be combined to solve problems. After defining a unit length, Descartes demonstrated procedures for adding, subtracting, multiplying, and dividing line segments and for graphically determining roots of equations. However, Descartes' *Géométrie* was difficult to understand and follow. It was written in French, not the language of scholarly communication at the time, and Descartes' writing style was often obscure in its meaning. In 1649, Frans van Schooten, a Dutch mathematician, published a Latin translation of Descartes' *Geometry*, adding his own clarifying explanations and commentaries. Descartes' friend and colleague, Florimond de Beaune, contributed an introduction to this edition. Van Schooten published a second edition in 1659, providing even more explanation and commentary. Now the message of Descartes' *Geometry* was available to a large reading audience, and it became an influential work, spurring the development of analytic geometry.

Unlike the *Regulae*, the first letters of the alphabet are used to designate indeterminate quantities, while the last (especially x, y, and sometimes z) to indicate the unknown. Solving Pappus' problem (in Book 1) is a perfect example of this new use of symbols. Descartes' mathematical writing was much more flexible and dynamic than the previous ones.

By introducing a series of important symbolic innovations, the *Géométrie* marked a definitive break with previous rhetorical writings; this work offers us "a flexible, almost modern symbolic writing, structured in clearly distinct propositions, and whose rhetorical commentary is reduced or non-existent" (Serfati 2005, 242). As mentioned, one of the major innovations introduced by Cartesian writing is the writing of power, which we still use today. The importance of this innovation was underestimated even by Descartes, who instead saw it only as a convenience. It would be precisely by developing Descartes' work on the exponent that Leibniz and Newton would discover new mathematical objects, as testified by the *Epistola Prior* and the *Epistola Posterior* of Newton, and the letter of Leibniz to Tschirnhaus of May 1678 in which the notion of transcendence in the mathematical sense appears for the first time. But power is only one possible example. Cartesian writing developed a game of combinations and substitutions, transformations, and interpretations that no other symbolic had ever done before. On the basis of a few rules, the symbolic developed by itself, transforming itself, and this led mathematicians to ask themselves new questions and discover new things. As said at the end of the previous section, the symbolic revolution brings to light the diagrammatic nature of writing.

Leibniz's design of *ars combinatoria* stemmed from these ideas of Descartes'. It is first of all a new way of understanding mathematical writing. As Couturat writes, "What is ultimately this combinatorics to which Leibniz also subordinates algebra? It is the general science of forms and formulas" (1901, 288; my translation). Leibniz developed a notation for the differential and integral calculus, the notation still used today, made it easy to do complicated calculations with little thought. "It was as though the notation did the work" (Davis 2000, 4). In this, he took Cartesian work to the extreme.

On one of his first trips to London, Leibniz presented the design of a machine capable of calculating by itself. This calculating machine was able to carry out the four basic operations of arithmetic. "Although Pascal had designed a machine that could add and subtract, Leibniz's was the first that could multiply and divide as well" (Davis 2000, 5). This machine incorporated an ingenious gadget that became known as a "Leibniz wheel." Calculating machines continued to be built incorporating this device well into the twentieth century.

However, in Leibniz's view, something similar could be done for the whole scope of human knowledge. For this reason, he developed the project of the universal characteristic, in which each symbol represented some definite idea in a natural and appropriate way. Leibniz thought of an encyclopedic compilation, "of a universal artificial mathematical language in which each facet of knowledge could be expressed, of calculational rules which would reveal all the logical interrelationships among these propositions" (Davis 2000, 5). Machines could be capable of carrying out calculations, freeing the mind for creative thought. Even with his optimism, Leibniz knew that the task of transforming this dream to reality was not something a man could accomplish alone. He did believe that a small number of capable people working together in a scientific academy could accomplish much of it in a few years. The whole human experience could be translated into characters and then developed in a clear and distinct way, according to mechanical rules. As he wrote in the letter to Jean Gallois, dated December 1678:

> I am convinced more and more of the utility and reality of this general science, and I see that very few people have understood its extent. [...] This characteristic consists of a certain script or language [...] that perfectly represents the relationships between our thoughts. The characters would be quite different from what has been imagined up to now. Because one has forgotten the principle that the characters of this script should serve invention and judgment as in algebra and arithmetic. This script will have great advantages; among others, there is one that seems particularly important to me. This is that it will be impossible to write, using these characters, chimerical notions such as suggest themselves to us. An ignoramus will not be able to use it, or, in striving to do so, he himself will become erudite. (translation from Davis 2000, 12)

Although we are still a long way from the Turing machine and von Neumann's architecture, or from Boole's logic, there was an important starting point in Leibniz's idea of a symbolic capable of dominating human knowledge.

## The Symbolic Revolution: Part 2

> There is nothing natural about software or any science of software. Programs exist only because we write them, we write them only because we have built computers on which to run them, and the programs we write ultimately reflect the structures of those computers. Computers are artifacts, pro-

> grams are artifacts, and models of the world created by programs are artifacts. Hence, any science about any of these must be a science of a world of our own making rather than of a world presented to us by nature. What makes it both challenging and intriguing is that those two worlds meet in the physical computer, which enacts a program in the world. Their encounter has posed new and difficult epistemological questions concerning what we can know both about the workings of the models and about the relation of the models to the phenomena they purport to represent or simulate. Answers to those questions would seem to depend, at least in part, on understanding programs as dynamic systems. (Mahoney 2002, 1)

The history of software does not coincide with that of computers. Although the birth and development of software have been a crucial moment in time, the history of the computer is much older and more complex. Initially considered a monotonous activity and of little importance with respect to the construction and development of hardware, software has become increasingly central, reaching predominance in the contemporary period. According to Mahoney (2005, 3), "perhaps for that reason, the history of software has so far either remained close to the machine or has stepped back altogether to view software in the large and as an object of commerce." Historians "have scarcely scratched the surface of applications software, the software that actually gets things done outside the world of the computer itself. [...] we have no studies of the software industry, both commercial and personal; of programming as a labor process and nascent profession; or of software engineering and its origins in the (continuing) software crisis" (3). Furthermore, "software will look more human when we take seriously the difficulties of designing and building it and undertake critical accounts of some of its failures" (5).

Just as the press has transformed the production and use of books and language (especially in the case of mathematics), so has software transformed the production and use of computers, also allowing the emergence of new professionals. In both of these processes, writing is the real protagonist. Two technologies (books and computers) take up and transform a third (writing). These two historical movements have replaced a world of objects with a world of texts, that is, a world in which objects are made of text or writing: this is the digital world. In the digital world, every object is composed of strings of code, that is, writing. Under a very thin layer of "skin," millions of lines of alphanumeric codes are hidden. Between these two historical movements (print and software), writing has become increasingly independent of meaning and human intervention and increasingly capable of intervening in reality and changing it. The materiality of the writing has taken over, but the meaning has not been erased; it has been transformed—indeed, the materiality of the writing has produced new meanings. These considerations preface the thesis of the next chapter: software as transcendental of the digital dimension.

The aim of this section is not to reconstruct an exhaustive history of computing and software: there have been many important studies on this topic. My intent is to define a general framework and highlight the importance of software in the history of writing. My thesis is that software represents a stage in the history of writing and that understanding this provides us with important elements with which to better understand the very nature of software. I am not saying that software is simply a particular form of writing. Software—as I tried to show in the previous chapters—is the result of the convergence of a range of phenomena: the development of mathematical logic and computability theory, the study of formal languages, and the development of electronic engineering and communication theory. As I will show in the next chapter, software was born from a creative synthesis of all these elements. It is a very interesting philosophical object precisely because of its multifaceted character.

## The Beginning

The year 1936 can be considered one of the most important turning points in the history of humanity. In that year, Turing's article "On Computable Numbers" was published in the *Proceedings of the London Mathematical Society*. This turning point can be summed up in three closely related key points. The first is the creation of two previously unknown mathematical concepts: those of machines and programs, "perhaps the most powerful mathematical objects ever invented" (Russell 2019, 33). The second point is the idea of universality or the connection between computation and the idea of a universal machine. "Universality means that we do not need separate machines for arithmetic, machine translation, chess, speech understanding, or animation: one machine does it all" (32). In other words, we can build a single machine capable of solving all solvable, that is, computable, problems. A machine is universal if it can simulate any computation possible. From a philosophical point of view, Turing (1936) equated agency and computability. Computability has become a universal ontological category. In other words, being is computable, that is, constructible, starting from some fundamental operations and elements. The third point concerns the limits of computation. Turing was able to understand even the limits of his own discovery: that some problems do remain undecided. The problem may be well defined, and there may also be a solution, but there is no procedure through which to find it.

Turing was interested in the definition of computable numbers and was concerned primarily with machines that produced unending sequences of 0s and 1s. These sequences were interpreted as binary representations of real numbers between 0 and 1. Each machine could compute a single well-defined sequence of 0 and 1. Turing described these machines using a notation known as "machine tables." "A table does not describe the physical structure of the machine, however, but its behaviour, or the sequence of basic operations that it will carry out" (Priestley 2011, 79). The simplest form of the machine table was a set of rows, each defining what the machine does in a single computational step. The behavior of a machine is determined by its configuration: each row in a table corresponds to a single machine configuration.

Although it represents the starting point of the modern history of computers and programming, the Turing (1936) machine is still only a mathematical concept. It is deeply connected to the developments of mathematical logic and, above all, the theory of formal languages—the work of Gödel. As Priestley (2011) writes:

> Turing wanted to take seriously the idea of computation by machines as a basis for an analysis of computability, and his problem was how to bridge the gap between the existing discipline of mathematical logic and the more concrete world of machines. He achieved this by the introduction of the machine table notation, which provided textual equivalents of potentially

> physical machines. By treating machine tables as texts in a formal language, it became possible for Turing to apply the well-developed resources of formal logic to the study of machines. (96)

There was a need for a second crucial turning point in order to arrive at the modern idea of computers. This second breakthrough, called ENIAC, was designed and built at the Moore School of Electrical Engineering under the leadership of John von Neumann and presented in February 1946. Campbell-Kelly et al. (2013), Ceruzzi (2003), and Mahoney agreed to considering ENIAC as the beginning of the history of the computer as we know it today. "It is really only in von Neumann's collaboration with the ENIAC team that two quite separate historical strands come together: the effort to achieve high-speed, high-precision, automatic calculation and the effort to design a logic machine capable of significant reasoning" (Mahoney 1988, 116). Priestley (2011) also spoke of "crucial innovation on the way to the development of the 'computer as we know it'" (124). ENIAC was the first computer based on the stored-program principle, which was then expanded into its successor, the UNIVAC. The innovation of the stored-program principle "led to the establishment of 'programming' (later 'software') as something both separate from and as important as hardware design" (Ceruzzi 2003, 21). This principle "remained remarkably stable during the evolution of computing from 1945 to 1995. Only toward the end of this period do we encounter significant deviations for it [...]." As mentioned above, a key role in the development of ENIAC was that of John von Neumann, author of the famous *First Draft of a Report on the EDVAC* (1945), which defined the fundamental points of what is now eponymously called the "von Neumann architecture."

The stored-program principle is pretty simple. It all stems from an initial defect of the ENIAC, which had to be manually reconfigured whenever a problem had to be solved. The reconfiguration was a long and tiring process. Coding was external to the machine; that is, the instructions were given to the machine through a process that was completely separate from it. There was a need to find an easier way to program and reprogram the machine in order to make it more flexible. Von Neumann's idea—this attribution has been disputed as many others had evoked this precise technical possibility (see Primiero 2020, 137)—was to store data and instructions in the same memory. This principle ensured that the instructions were always at hand and could be changed at great speed. Furthermore, as the instructions were stored in the memory, they were immediately accessible to the machine, which could therefore use them in a versatile manner. Most likely, as Priestley (2011) pointed out, von Neumann's idea was inspired by the 1936 Turing paper:

> On the one hand, von Neumann was aware of and admired Turing's work, and would certainly have been familiar with the design of the universal machine. On the other hand, there is no explicit mention of Turing in the *Draft Report* which seemed more concerned to link the new machine with the developing area of cybernetics than directly to logic. (139)

We have to avoid the mistake of conflating the stored-program principle with the entire von Neumann architecture. The draft was written by von Neumann for the development of the successor to ENIAC, the EDVAC, which was built at the Ballistic Research Laboratory in 1949. This computer is characterized by three elements: (a) the stored-program concept, (b) the hardware structure, and (c) the sequentialization of programs with all memory values transferred through the arithmetical unit (Primiero 2020, 131). In the draft, von Neumann "focuses on the logical control of a very high automatic digital computing system defined as a device to carry out instructions to perform highly complex calculations. Instructions reflect the definiteness, clarity, and exhaustiveness required by algorithmic processes" (131). With this goal in mind, von Neumann described four key elements of the EDVAC architecture, which would later become common to all computers: (a) the central arithmetic unit (it performs basic arithmetic operations), (b) the central control unit (the logical unit), (c) the memory (it stores all program instructions and data), and (d) the recording unit (the input-output device).

Neumann's draft defined a paradigm and had enormous importance in the history of the computer. However, it does not represent the origin of software. The process from the definition of the stored-program principle to the development of the first programming languages was long and elaborate. As Mahoney (1990, 2) recalled, "the computer emerged as the joint product of electrical engineering and theoretical mathematics and was shared by those two groups of practitioners, whose expertise intersected in the machine and overlapped on the instruction set. Both groups apparently looked upon programming as incidental to their respective concerns. [...] As long as the computer remained essentially a scientific instrument, [...] programming remained relatively unproblematic."

## The Turning Point

The fundamental turning point in the history of software took place in the early 1950s with the emergence of the first automatic programming methods. The term "coding" was used to refer to the process of translating the instructions of a program into the coded form used inside the machine. "In some cases this was carried out entirely by hand, but following the example of the EDSAC, many installations devised a set of 'initial orders' which would translate instructions from a more human-friendly form into machine code" (Priestley 2011, 18). The Electronic Delay Storage Automatic Calculator (EDSAC) was one of the first digital computers. Developed by Wilkes between 1947 and 1949, it was based on the von Neumann architecture. Although there were some automatic processes in translating operations into machine code, much of the programming was still done by hand.

Automatic programming methods deeply transformed the programmer's work by making it less time-consuming, repetitive, and mechanical. Furthermore, these methods improved the ability to find errors in the code, thus making the whole process faster and more reliable. The first example of automatic programming was the A-0, developed by Hopper between 1951 and 1952 and used on the UNIVAC, the first commercial computer created in the United States. The A-0 represented a decisive step forward compared to the subroutine method (a short program that could be recalled, even several times, from the main program to solve a specific and frequent problem, thus facilitating programming and saving memory), which was introduced a few years earlier by Wilkes, Wheeler, and Gill. The idea of subroutines was already present in Turing's project for the ACE computer:

> We also wish to be able to arrange for the splitting up of operations into subsidiary operations. This should be done in such a way that once we have written down how an operation is to be done we can use it as a subsidiary to any other operation. (Turing and Woodger 1986, 34)

Hopper defined the A-0 as a "compiler," that is, "a programming-making routine which produces a specific program for a particular problem" (Ceruzzi 2003, 85). However, what Hopper had in mind was not exactly what is meant today as a compiler. For Hopper, "a compiler handled subroutines stored in libraries"; it was "a program that copied the subroutine code into the proper place in the main program where a programmer wanted to use it" (85). The first compiler in the modern sense was developed by Laning and Zieler at MIT for the Whirlwind computer. The program "took as its input commands entered by a user, and generated as output fresh and novel machine code, which not only executed those commands, but also kept track of storage locations, handled repetitive loops, and did other housekeeping chores" (86). However, Laning and Zieler's compiler was limited to one purpose: solving mathematical equations. Just as the printers introduced a new technology to fix the defects of the works of amanuenses, so did Hopper and others introduce a new technology to fix the defects of the first programming, which was a very long, mechanical, and fully manual job.

The first real compilers and programming languages were born in the late 1950s. The most classic example was FORTRAN (Formula Translation), which was created between 1954 and 1957, by a group led by John Backus, and subsequently expanded with new versions starting from 1958. FORTRAN was used for the IBM 704, essentially for academic purposes—mathematical and logical in particular. It was hugely successful until the 1990s. However, FORTRAN was not machine-independent in the sense that it was still only meant to be used on a single machine; using it on another machine (even of the same type) involved a huge waste of money and effort. The first example of machine-independent programming language was the Common Business Oriented Language (COBOL). Designed in 1959 and officially launched on the market in 1961, COBOL was aimed primarily at the business world: "the goal was to develop a programming language specifically aimed at the needs of the business data processing community" (Ensmenger 2010, 93). The project was led by Hopper and supported by the US Department of Defense. COBOL was a more readable language than FORTRAN, less technical and oriented more toward practical applications. As mentioned above, it was machine-independent in the sense that it could be used on several machines. "COBOL became one of the first languages to be standardized to a point where the same program could run on different computers from different vendors and produce the same results" (Ceruzzi 2003, 92). It was applauded by the corporate world for two reasons: (a) it was non-technical, so it could be understood and used even by non-programmers, which allowed cost-cutting by avoiding new staff hires; (b) it had more practical uses. COBOL received heavy criticism from academics, who preferred FORTRAN because it was considered more rigorous.

The third peak of the first era of software was ALGOL (algorithmic language), which was developed mainly in Europe between 1958 and 1960. ALGOL was the result of the work of a committee composed of academics, professional programmers, and users. It was an international project aimed at better defining programming standards. It was also a very rigorous and aesthetically beautiful language and could be considered "a remarkable achievement in the nascent discipline of computer science" (Ensmenger 2010, 104). The desire was for ALGOL to combine two dimensions: beauty and rigor. This language introduced important innovations from a syntactic point of view and a way of representing algorithms that would become the model for all subsequent languages. Nevertheless, its diffusion was very limited in the United States, where it was considered an intellectual curiosity rather than a program with practical applications (104), although it was the basis of another important language, PASCAL, which had a huge spread in the 1970s.

Despite the birth and diffusion of these first programming languages (there are many more examples), between the years 1950 and 1960, the world of software was undergoing a deep crisis. On one hand, there was a growing demand for programmers and computer scientists because of the progressive spread and commercialization of computers. On the other hand, due to the increase in software costs and its centrality in the management of complex processes, there was the need for a greater professionalization of programming and, therefore, the creation of curricula, courses of qualification, and certifications that could define the skill set and abilities needed to become a programmer. Managers, aware that software was becoming a crucial business asset, wanted programmers to be more integrated into the business process and more reliable. For many managers, computer boys were excellent technicians, albeit unprofessional. The programmers were considered a "double-edged sword." Increasing the professionalism of programmers meant being able to solve many managerial problems in a short time.

Above all, there was the need for clarity. Was it not clear what computer science and software were—engineering, mathematics, psychology, linguistics, business management, or whatever? What skills were needed to become a programmer? What exactly was the role of the programmer? There was no agreement on the training needed to reliably carry out the profession. This debate reflected "the deep intellectual and ideological schisms that existed within the programming community" (Ensmenger 2010, 168). Some believed that programming was, above all, an art and, therefore, that it was based on a personal nature that could not be defined. This idea also fueled a certain type of behavior in the world of programmers: arrogance, eccentricity, asociality, a spirit of independence, and so on. Conversely, others believed that programming was only technical and that it should be defined in relation to the concrete purpose to which it was applied. According to this point of view, it was necessary to introduce clear standards, associations, and ethical codes to allow better use of programmers in companies. "Computer scientists expressed disdain for professional programmers, and professional programmers responded by accusing computer science of being overly abstract and irrelevant" (129). There were also weightier criticisms from managers, with many believing that software was overly expensive, useless, and "an unprofitable morass."

This first art–business schism was connected to another, that between the world of corporations and universities, which was then reflected in the clash between two great associations: the Association for Computing Machinery (ACM; born in 1947, it was theoretically oriented and composed mainly of academics) and the Data Processing Management Association (DPMA; born in 1962, it was successor to the National Machine Accountants Association). "Although many practitioners agreed on the need for a programming profession, they disagreed sharply about what such a profession should look like. What was the purpose of the profession? Who should be allowed to participate" (168)? Academics struggled to define the status of a new discipline based on a body of shared knowledge. They wanted to turn technology into a science. In academia, the idea of "computer science" was highly contested: most academics saw it as nothing more than a branch of electronic engineering, applied mathematics, or statistics and that, therefore, it should also be assimilated at an institutional level in these disciplines. "That was a real fear within the nascent computer science community that its discipline was not being taken seriously, that it was considered by many little more than a momentary aberration in the field of mathematics and electrical engineering" (116).

Nonetheless, business programmers worked to improve the reputation and role of their business in the corporate hierarchy and fought against being considered mere technicians like so many others. The issue of what programming was—as an intellectual and occupational activity—and where it fit into traditional social, academic, and professional hierarchies "was actively negotiated during the decades of the 1950s and 1960s" (169). Programmers were aware of the fragility of their role, and for this reason, they tried to defend it by claiming the originality of their profession.

The answers to this crisis were manifold. For example, the DPMA instituted the CDP examination, a certification guaranteeing a basic level of skills (English, mathematics, statistics, etc.) for all programmers. Companies saw in it a way to ensure consistency and quality in programming work. The CDP achieved great success in the 1960s and 1970s, before being completely reformed following the denunciation of several scandals and irregularities. Its credibility progressively diminished. Another example was the establishment of the American Federation of Information Processing Societies (AFIPS) in 1975. This association was born to create unity among the various professional associations of US programmers, but it soon proved to be a failure: "It was crippled by a weak charter and a lack of a tangible support from its founding societies" (188). The DPMA did not join it until 1974, albeit without great enthusiasm. Individual programmer membership was also scarce. Furthermore, there were numerous tensions between the various associations. "The reasons behind the failure [of these associations] suggest the limitations of professional associations as an institutional solution to the software crisis" (189). Various groups of programmers shared the need to bestow a stronger identity onto the profession; however, they had completely different ideas about what this identity should be. The associations fueled this debate without offering credible solutions to companies, so they automatically lost value. In the late 1960s, programmers were seen not as professionals but as mere technicians: they did not appear to be representatives of a well-defined discipline based on a solid body of knowledge.

The second answer to the software crisis was the theoretical definition of computer science. A decisive step in this direction was taken by Donald Knuth, a computer scientist at Stanford University who, in 1968, published the first volume of a work that would become canonical: *The Art of Computer Programming*. Knuth first introduced the idea that the algorithm was the fundamental object of computer science. In this work, he not only offered a first history of computing by linking it to the history of science but also defined a series of methods based on the notion of the algorithm as the essence of intelligence. Computer science was not, according to Knuth, the study of computers; it was the science of algorithms. The centrality of the algorithm also provided the nascent computer science with a practical agenda: the study of algorithms proved to be very productive. "Knuth provided a theoretical basis for computing that was practical, and his book established it as the algorithm, a formal procedure that one can use to solve a problem in a reasonable length of time, given the constraints of actual computing machine" (Ceruzzi 2003, 103). This methodological choice was also confirmed by the ACM, which included the study of algorithms in its official curriculum in 1968; this is "a landmark moment in the history of the discipline" (Ensmenger 2010, 133). Computer science claimed to be a field of study and research with a well-identified knowledge base and methods. The ACM curriculum was adopted by many universities; however, it was strongly criticized by corporations, which considered that it was overly abstract, useless, and completely detached from the concrete problems to which the software had to be applied.

The third answer to the software crisis was the NATO Conference on Software Engineering in October 1968 in Garmisch. It was attended by academics, computer scientists, managers, and military personnel. As stated at the beginning of the official conference report:

> The Conference was attended by more than fifty people, from eleven different countries, all concerned professionally with software, either as users, manufacturers, or teachers at universities. The discussions cover all aspects of software including relation of software to the hardware of computers; design of software; production, or implementation of software; distribution of software; service on software. [...] Although much of the discussions were of a detailed technical nature, the report also contains sections reporting on discussions which will be of interest to a much wider audience. This holds for subjects like the problems of achieving sufficient reliability in the data systems which are becoming increasingly integrated into the central activities of modern society; the difficulties of meeting schedules and specifications on large software projects; the education of software (or data systems) engineers; the highly controversial question of whether software should be priced separately from hardware. (NATO 1968, 4)

The conference mainly responded to a technical need: many software projects were too complex, difficult to manage, too expensive, too long, or often managed by unprofessional people. Furthermore, "the increasing size of software projects introduced two new elements into programming: separation of design from implementation and management of programmers" (Mahoney 1990, 2). This required a transformation of the methods and concepts that were the basis of programming. Therefore, the conference represented an important moment of reflection, especially in terms of the relationship between software and the business world. "With the Garmisch conference, software began to be conceptualized as a problem—and the 'software crisis' was constituted as a point of origin for the discipline of software engineering" (Frabetti 2014, 57). It was necessary to adopt a more methodical approach, in line with the industry, to define an object and a method. "The general consensus among historians and practitioners alike is that the Garmisch meeting marked a major cultural shift in the perception of programming" (Ensmenger 2010, 196). Thus, the expression "software engineering" was coined, although it did not characterize "an ongoing activity but rather to express a desire for one" (Mahoney 1990, 1).

The Garmisch conference marked the transition from programming as an almost secret art to programming as a rigorous discipline based on methods in engineering and management. As one of the participants, Alexander D'Agapeyeff, pointed out, "programming is still too much of an artistic endeavour. We need a more substantial basis to be taught and monitored in practice on the: (i) structure of programs and the flow of their execution; (ii) shaping of modules and an environment for their testing; (iii) simulation of run time conditions" (NATO 1968, 34). Another participant complained of the severe lack of competence: "I think that one of the major problems with the very large programming projects has been the lack of competence in programming which one observes as soon as one goes above the very bottom level. People begin to talk in vague general terms using words like 'module', and very rarely ever get down to the detail of what a module actually is. They use phrases like 'communicate across modules by going up and then down'—all of the vague administratese which in a sense must cover up a native and total incompetence" (NATO 1968, 49).

A crucial problem for many of the conference attendees was the pace of growth of software systems, their complexity, and social importance. According to Ascher Opler (IBM), "I am concerned about the current growth of systems, and what I expect is probably an exponential growth of errors. Should we have systems of this size and complexity? Is it the manufacturer's fault for producing them or the user's for demanding them? One shouldn't ask for large systems and then complain about their largeness" (NATO 1968, 54). This case shows the distinct emergence of two positions: one favoring management, and therefore the control of the working process, and the other focusing more on research and development. "Even those proposals that seemed to be most explicitly technical, such as those advocating structured programming techniques or high-level language developments, contained a strong managerial component" (Ensmenger 2010, 198). However, the conference did not solve the most fundamental problems; there were persistent tensions and differences between the participants, who came from different fields and were of different backgrounds. "Not only do the conference participants feel the pressure of social demands on them; they also feel that software development reaches its point of crisis when society pushes the boundaries of state-of-the-art technology. But do these demands come from society or from technology itself?" (Frabetti 2014, 58).

The Garmisch conference represents a historic passage in which two very important events occurred for the first time: (a) the awareness of the software crisis and the need for an effective solution was expressed; (b) a serious discussion on the future of software was initiated in view of an effective, practical, and theoretical foundation of the discipline. The world of academia and that of industry reached a compromise and began collaborating. The subsequent Rome conference in 1969 was a failure: attendees criticized it as boring and conflict-ridden. There was no longer the Garmisch atmosphere of development and rebirth. However, the problems, the attendees, and the available technologies were the same. According to Ensmenger (2010), the real difference between the two conferences lay in the fact that, in the first, the participants found a common point in the sense of an obvious emergence, even if vague. In Garmisch, the solution was widely shared by all groups of programmers; the concept of "software engineering" was sufficiently fluid to be adopted by anyone interested in programming. In Rome, this atmosphere changed, and old rivalries between groups of programmers resurfaced.

> Unlike the first conference, at which it was fully accepted that the term software engineering expressed a need rather than a reality, in Rome there was a tendency on the part of some participants to talk as if the subject already existed. An even greater contrast to the Garmisch conference was the gulf that opened up between the researchers and practitioners, and the ill-tempered nature of many of the discussions. And it became clear during the conference that the organizers had a hidden agenda, namely that of persuading NATO to fund the setting up of an International Software Engineering Institute. However things did not go according to their plan. (Randell 2018, 4)

The Garmisch conference did not eliminate the tensions at all; rather, it translated them into a very different context—a different awareness. The term "software engineering" was more indicative of the need than a reality. The dispute over the status of this alleged discipline was ongoing, as demonstrated in the difference between the position of Sommerville:

> Software engineering is an engineering discipline that is concerned with all aspects of software production from the early stages of system specification through to maintaining the system after it has gone into use. (Sommerville 2016, 23)

And that of Parnas:

> The most accurate title for the people that are now called "Software Engineers" would be "Paid Software Developers". They are professionals in the sense that they are paid for what they do, which is develop software but they have no credentials that are relevant to software development, the vast majority of the ones who are Professional Engineers gained that credential in another discipline and then migrated into software development without significant education about software design… When the term "Software Engineering" was introduced, many asked a simple question. "How is "Software Engineering different from programming"? (Parnas 2011, 12)

The software crisis never ended. Software itself, due to its heterogeneous nature—ambiguity regarding the very activity of programming and the profession connected to it—made this crisis inevitable. This is the central thesis of Ensmenger's (2010) book: software is perpetually in crisis, despite being one of the most powerful and profitable industries in modern history. "Despite their seemingly intangible nature, software applications are always inextricably linked to a network of social and technological systems" (226). The history of software is the history of an incredible technological success. However, the identity of software continues to be marked by the perception of a perennial crisis or failure. In terms of actor-network theory (Latour 2015), software can be defined as a network of actors in constant interaction with each other, based on different power relationships. This makes the ontology of software an even more complex task than indicated in the first chapter of this book. Following Latour's terminology, software can be seen as an expanding *collective*, a network of actors assimilating everything. Everything, today, exists only if it is "accepted" in this collective, "translated" in its terms, that is, into data and code strings, and therefore "maintained" and "defended." According to Ensmenger, the contradiction of software (success and crisis) can only be understood if we look at the fact that software connects the computer to the human world, and therefore, the perception of crisis that characterizes it does not concern its functionality but, rather, the social dynamics that it reflects. For this reason, according to Ensmenger, any solution to the software crisis must reflect this heterogeneity.

## Conclusions

Why is software connected to writing and in particular to mathematical writing? To say that software is writing is not to say a triviality. Proust's *Recherche du temps perdu* is also writing, but it is writing in a very different way from software. For the *Recherche*, writing is a means of creating meaning in the minds of its readers—as Ricoeur shows in the second volume of *Time and Narrative*. For a program, writing is not just a means. In the modelization phase, we use a mathematical language to build the algorithm. This mathematical language is then translated into a series of machine operations. However, the CPU does not know the meaning of those operations; it does not interpret the meaning of the process. The CPU does not know anything about math but does math. The machine does only one thing: it re-writes and stores the code. Performing an operation means re-writing a piece of code and storing it—that is, re-writing it again—in a place in its memory. *Re-writing the code translates into action*. The transition from code to action is immediate. There is no meaning. In order to work, software *must eliminate* the meaning and reduce itself to a pure process of writing and re-writing—a circulation and transformation of electrical signals. From this point of view, software takes the revolution begun by Descartes to its extreme consequences. The analysis of the symbolic revolution allows us to understand why and how software is based on mathematics and writing, and in what sense it eliminates meaning. What happens in the transition from Cardan to Descartes? In mathematical writing, the presence of interpretation is displaced. The interpretation is shifted from start to finish. It does not condition writing but develops from it. The mathematician relies on the power of writing and combinatorics as heuristic tools. Thus, software is writing because a regression to writing occurs in it.

Now, analyzing Proust's *Recherche* in *Time and Narrative*, Ricoeur distinguishes three moments: the prefiguration of the action, the configuration of the text, and the refiguration—as we saw before. His method combines literary criticism with philosophical reflection, and in particular with ontology—in a non-metaphysical sense. If we want to use this method also in the case of software, we must use critical code studies in a hermeneutic sense, that is, as a premise and condition of ontological inquiry. For example, the poetics of the game defined by Wardrip-Fruin (2020), starting from the analysis of Pac-man, can be the starting point for reinterpreting the being of the game in the human world. Critical code studies can provide us with a mediation between code and the action of code in the human world.

## References

Cajori, F. 1993. *A History of Mathematical Notations*. New York: Dover Publications.

Campbell-Kelly, M., W. Aspray, N. Ensmenger, and J. Yost. 2013. *Computer: A History of the Information Machine*. London: Routledge.

Ceruzzi, P. 2003. *A History of Modern Computing*. Cambridge, MA: MIT Press.

Couturat, L. 1901. *La logique de Leibniz*. Paris: Alcan.

Davis, M. 2000. *The Universal Computer: The Road from Leibniz to Turing*. London & New York: Taylor & Francis.

Descartes, R. 1954. *The Geometry of René Descartes (with a Facsimile of the First Edition)*. London: Dover Publications.

———. 1988. *Œuvres Philosophiques*. Edited by F. Alquier. Paris: Garnier.

———. 2018. *Rules for the Direction of the Mind*. London: Franklin Pub.

Eisenstein, E. 1983. *The Printing Revolution in Early Modern Europe*. Cambridge University Press.

Ensmenger, N. 2010. *The Computer Boys Take Over*. Cambridge, MA: MIT Press.

Frabetti, F. 2014. *Software Theory*. London: Rowman & Littlefield.

Havelock, E. 1963. *Preface to Plato*. Cambridge: Cambridge University Press.

Heath, T.L. 1981. *History of Greek Mathematics*. New York: Dover Publications.

Ihde, D. 1990. *Technology and The Lifeworld*. Bloomington: Indiana University Press.

Israel, G. 1997. The Analytical Method in Descartes' *Géométrie*. In *Analysis and Synthesis in Mathematics: History and Philosophy*, ed. M. Otte and M. Panza, 3–34. Dordrecht: Kluwer Academic Publishers.

Jullien, V. 1996. *Descartes, La Géometrie de 1637*. Paris: Puf.

Katz, V.J. 2009. *A History of Mathematics*. London: Pearson.

Knauff, M. 2013. *Space to Reason: A Spatial Theory of Human Thought*. Cambridge, MA: MIT Press.

Latour, B. 2015. *Reassembling the Social*. Oxford University Press.

Mahoney, M. 1988. The History of Computing in the History of Technology. *Annals of the History of Computing* 10: 113–125.

———. 1990. The Roots of Software Engineering. *CWI Quarterly* 3 (4): 325–334.

———. 2002. Software as Science—Science as Software, and Probing the Elephant: How do the Parts Fit Together? In *History of Computing: Software Issues*, ed. Ulf Hashagen, Reinhard Keil-Slawik, and Arthur Norberg. Berlin: Springer Verlag; presented at the International Conference on the History of Computing 2000 at the Heinz Nixdorf Museums Forum, Paderborn, Germany, 5–7 April 2000.

———. 2005. 'The History of Computing(s)' a Lecture in the Series 'Digital Scholarship, Digital Culture', at the Centre for Computing in the Humanities, King's College, London, 18 March 2004; published version in *Interdisciplinary Science Reviews* 30 (2).

Manders, K. 2006. Algebra in Roth, Faulhaber, and Descartes. *Historia Mathematica* 2 (33): 184–209.

McLuhan, M. 1962. *The Gutenberg Galaxy: The Making of Typographic Man*. University of Toronto Press.

NATO. 1968. *Software Engineering: Report on a Conference Sponsored by the Nato Science Committee* (Garmisch, Germany, 7–11th October).

Ong, W. 1982. *Orality and Literacy: The Technologizing of the Word*. London & New York: Routledge.

———. 2002. *The Ong Reader. Challenges for Further Inquiry*. New York: Hampton Press.

Parnas, D.L. 2011. Software Engineering: Multi-person Development of Multi-version Programs. In *Dependable and Historic Computing*, ed. C.B. Jones and J.L. Lloyd, 413–427. London: Springer.

Priestley, M. 2011. *A Science of Operations. Machine, Logic and the Invention of Programming*. London: Springer.

Primiero, G. 2020. *On the Foundations of Computing*. Oxford: Oxford University Press.

Rabouin, D. 2010. *Mathesis universalis: l'idée de "mathématique universelle" d'Aristote à Descartes*. Paris: Puf.

Randell, B. 2018. Fifty Years of Software Engineering. https://arxiv.org/pdf/1805.02742.pdf.

Russell, S. 2019. *Human Compatible. AI and the Problem of Control*. New York: Random.

Serfati, M. 2005. *La revolution symbolique. La constitution de l'écriture symbolique mathématique*. Paris: Pétra.

Sommerville, I. 2016. *Software Engineering*. London: Pearson.

Turing, A. 1936. On Computable Numbers, with an Application to the Entscheidungsproblem. *Proceedings of the London Mathematical Society Volume* s2-42: 230–265

Turing, A. and Woodger, M. 1986. A M. *Turing's ACE Report of 1946 and Other Papers*. Cambridge, MA: MIT Press.

Wardrip-Fruin, N. 2020. *How Pac-Man Eats*. Cambridge, MA: MIT Press.

Wolf, M. 2007. *Proust and the Squid*. London: Icon Books.

———. 2018. *Reader, Come Home. The Reading Brain in a Digital World*. London: Harper Collins.

# 5

# The Digital Reflective Judgment

## Introduction

As Mahoney (2005, 6) writes, "programs are artifacts, and we must learn to read them as such. What makes it difficult is precisely that the representations are operative." Ultimately "it is their behavior rather than their structure (or the fit between structure and behavior) that interests us. We do not use computers by reading programs; we interact with programs running on computers" (Mahoney 2005, 6). With software, the nature and function of language change tremendously. There is no longer a writer and a reader, a sender and a receiver. The syntactic and semantic dimensions are unified in a single performative act. As I have tried to demonstrate in previous chapters, software is a process, not an object. It has a dynamic nature. "The primary source for the historian of software is the dynamic process, and, where it is still available, it requires special techniques of analysis. It is a dynamic artifact, a running process, expressed in a static artifact, a program, and the relation between the two is indeterminate" (Mahoney 2005, 6). In this chapter, I focus on the

This chapter is a new development of Possati (2020).



L. M. Possati, *Software as Hermeneutics*,
https://doi.org/10.1007/978-3-030-63610-4_5

nature of this process, developing what was said in Chaps. 3 and 4. This chapter aims to unify the historical and theoretical parts of this book.

The central thesis of this chapter is that software is a specific form of reflective judgment, in Kantian terms. I call this type of reflective judgment "digital reflective judgment." In the first section of the chapter, I analyze Kant's concept of reflective judgment. In the second and third, I introduce the parallelism between software and reflective judgment. In the fourth section, I identify the transcendental structures at the root of digital reflective judgment. I call these structures "the graphic mind." In the sixth section, I introduce another transcendental condition of digital reflective judgment, namely the principle of finality: another parallel with the Kantian doctrine. The fifth section is an interlude in which I deal with the concept of writing starting from Bruno Latour's actor-network theory. This interlude is intended to compensate for the too subjectivist tendency of Kantian theory.

An important caveat. To argue that the "graphic mind" is the transcendental condition of software does not mean to deny that software is a formal language with a mathematical structure. The graphic mind—and digital reflective judgment in general—is what allows and guides the interpretation and application of that formal language. Between the formal structure of language and its material realization through the machine, there is a necessary mediation that allows the unity of the process. This mediation is the graphic mind, that is, a material transcendental condition, which lies outside the mind. A technological transcendental.

Moreover, as we saw in the previous chapter, the contradiction between the engineering dimension and the artistic dimension is part of the nature of software. The digital reflective judgment mediates between these two dimensions, and so explains this contradiction. This improves what I claimed in the previous parts of the book, especially about Ricoeur's approach to narrative. This is fundamental to understanding the hermeneutic dimension of software, that is, how to conciliate through software a program understood as a narrative and an algorithm understood as a pure calculation without semantics.

## The Kantian Reflective Judgment

According to Kant, a "judgment" (*Urteil*) is a specific kind of "cognition" (*Erkenntnis*), that is, a conscious mental representation of an object (*Critique of the Pure Reason*, A320/B376) [1] This representation has a synthetic form: it unifies and organizes raw, unstructured sensory data according to universal concepts, rules, or principles. Judgment is essentially the faculty of thinking of the "particular" (the representation of a singular thing) as being contained under the "universal" (the general representation). This synthesis is the characteristic output of the "power of judgment" (*Urteilskraft*). The power of judgment is a cognitive "capacity" (*Fähigkeit*), more specifically, a *spontaneous* and *innate* cognitive capacity. By virtue of this, it is the "faculty of judging" (*Vermögen zu urteilen*) (A69/B94), which is also the same as the "faculty of thinking" (*Vermögen zu denken*) (A81/B106). It is a controverted question whether, according to Kant, there is only one kind of synthesis or many different kinds. Moreover, terms such as "spontaneity" or "concept" can have different meanings in Kant's works. These issues are closely linked to the recent debate about Kant's conceptualism versus Kant's non-conceptualism in relation to his theory of judgment and the ensuing implications for interpreting and critically evaluating his transcendental idealism and the "Transcendental Deduction of the Pure Concepts of the Understanding" (see Hanna 2001, 2005, 2006, 2017; Land 2011, 2015, 2016; Ginsborg 2006). However, I do not want to tackle these issues here.

I want to stress three points. First, for Kant, judgments are essentially propositional cognitions—from which it immediately follows that rational humans are, more precisely, propositional animals. The connection between judgment and language is, therefore, essential.

Second, Kant distinguishes the logical form and the propositional content of a judgment. The logical forms are summarized in the "table of judgements" (*Critique of the Pure Reason*, A52–55/B76–79). The propositional contents, which are more fundamental than the logical forms, are classified according to two conceptual couples: a priori/a posteriori, analytical/synthetical.[2] Briefly, the propositional content of a judgment can vary along at least three dimensions: (1) its relation to sensory content,

(2) its relation to the truth-conditions of propositions, and (3) its relation to the conditions for objective validity.[3]

Third, Kant distinguishes between propositional contents and *the use of* propositional contents. It is possible for a rational subject to use the same propositional content in different ways. What does this mean? The fundamental difference is that between (a) theoretical use and (b) non-theoretical use. The first use aims to formulate true propositions about the world in order to obtain some knowledge, that is, science. The second use does not aim to formulate true propositions about the world; thus, its aim is pragmatic, moral, aesthetic, or teleological.

To specify the distinction between theoretical and non-theoretical uses of propositional contents, Kant introduces the distinction between "determining" judgment and "reflective" judgment. In the first "Introduction" to *The Critique of Judgment* (1790), he writes:

> Judgement in general is the faculty of thinking the particular as contained under the universal. If the universal (the rule, principle, or law) is given, then the judgement which subsumes the particular under it is determining. This is so even where such a judgement is transcendental and, as such, provides the conditions a priori in conformity with which alone the subsumption under that universal can be effected. If, however, only the particular is given and the universal has to be found for it, then the judgement is simply reflective. (Kant 2016, 53)

Here, Kant develops some remarks about the regulative use of the ideas of reason, which appeared in the first *Critique*'s *Appendix* to the "Transcendental Dialectic," in particular, the distinction between "apodictic" and "hypothetical" uses of judgment (A647/B675). The difference between the determining and reflective uses of a judgment has to do with the way in which the synthetical structure of the judgment is interpreted. The determining use presupposes a high-order representation under which to subsume the particular. It determines an individual or narrower concept by using a given general "determinable" concept or principle. The reflective use follows the opposite way. It presupposes a particular individual or narrower concept and advances from it toward a universal or more general concept. Thus, the reflective judgment directly invokes the cognitive subject's ability to form higher-order representations through the act of reflection (*Überlegung*) and, consequently, to be rationally self-conscious or apperceptive. The aesthetic judgment (the judgment of taste) and the teleological judgment are expressions of the reflective use of propositional contents. It is worth reading the entire passage from the first "Introduction" to *The Critique of Judgment*:

> The determining judgement determines under universal transcendental laws furnished by understanding and is subsumptive only; the law is marked out for it a priori, and it has no need to devise a law for its own guidance to enable it to subordinate the particular in nature to the universal. But there are such manifold forms of nature, so many modifications, as it were, of the universal transcendental concepts of nature, left undetermined by the laws furnished by pure understanding a priori as above mentioned, and for the reason that these laws only touch the general possibility of a nature (as an object of sense), that there must needs also be laws in this behalf. These laws, being empirical, may be contingent as far as the light of our understanding goes, but still, if they are to be called laws (as the concept of nature requires), they must be regarded as necessary on a principle, unknown though it be to us, of the unity of the manifold. The reflective judgement which is compelled to ascend from the particular in nature to the universal stands, therefore, in need of a principle. This principle it cannot borrow from experience, because what it has to do is to establish just the unity of all empirical principles under higher, though likewise empirical, principles, and thence the possibility of the systematic subordination of higher and lower. Such a transcendental principle, therefore, the reflective judgement can only give as a law from and to itself. It cannot derive it from any other quarter (as it would then be a determining judgement). Nor can it prescribe it to nature, for reflection on the laws of nature adjusts itself to nature, and not nature to the conditions according to which we strive to obtain a concept of it—a concept that is quite contingent in respect of these conditions. (Kant 2016, 23)

Aesthetic and teleological judgments can only be reflective: they do not produce knowledge.[4] While determining judgment is based on several a priori principles, that is, the principles of pure reason (transcendental logic), which are the basis of objective knowledge, reflective judgment has only one a priori transcendental principle—finality—which is universal and subjective at the same time. Therefore, reflective judgment interprets the particular case according to finality. It can have two forms: (1) in the first case, the principle of finality is applied to the relationship between the subject and the representation of the particular case and, therefore, to the spontaneous agreement between imagination and understanding, which produces a delight—interpreted as a sign of finality; (2) in the second case, finality is applied to the organization of nature through understanding and reason. In this case, it can only be subjective because it cannot be the object of a possible experience, that is, a phenomenon. Through the faculty of judgment, nature is represented as if a supreme intelligence had arranged the unity of all empirical laws.

However, the principle of finality is not to be confused with practical finality, as we can read in the first "Introduction" to *The Critique of Judgment*:

> [...] this transcendental concept of a finality of nature is neither a concept of nature nor of freedom, since it attributes nothing at all to the object, i.e., to nature, *but only represents the unique mode in which we must proceed in our reflection upon the objects of nature with a view to getting a thoroughly interconnected whole of experience, and so is a subjective principle, i.e., maxim, of judgement*. For this reason, too, just as if it were a lucky chance that favoured us, we are rejoiced (properly speaking, relieved of a want) where we meet with such systematic unity under merely empirical laws: although we must necessarily assume the presence of such a unity, apart from any ability on our part to apprehend or prove its existence. (Kant 2016, 46; emphasis added)

The principle of finality is the essence of the faculty of judgment as an autonomous faculty with respect to understanding and practical reason. In the principle of finality, the faculty of judgment gives itself a law in order to think about the unity of nature. The faculty of judgment presupposes this law in order to obtain an overall view of nature that is acceptable to us. Therefore, finality has a hypothetical nature. In *Logik* (§81), Kant describes finality as the "analogon" of the logical universality. Thus, the faculty of judgment produces inductive and analogical reasoning.[5]

Even if it does not have a cognitive function, reflective judgment plays a crucial role in science from a heuristic and methodological point of view. In the third *Critique*, Kant emphasizes the need for teleological judgment for the study of biology. If the understanding explains a coherent physical science based on universal laws, it is not sufficient to explain the smallest and simplest living organisms. The life of a worm or the growth of a blade of grass can never be understood starting from a determining judgment; it can only be understood through reflective judgment. Notions such as "gender" or "species" have a heuristic and methodological value to the extent that they are used in connection with determining judgment. They cannot be the basis of synthetic a priori judgments, but they can help in explaining what cannot be stated or formulated in synthetic a priori judgments.

## The Nature of the Digital Reflective Judgment

I propose that software be conceived from a Kantian perspective, that is, as a kind of reflective judgment. I call this new form of reflective judgment "digital reflective judgment" (henceforth DRJ). Why is it that software cannot be compared to determining judgment? It cannot be so compared because its function is not limited to subsuming a concrete problem to an abstract computational structure, namely the Turing machine. Software is not a mechanical process.

Let us look closely at *what the programmer does*. Like reflective judgment, software starts from an individual case, that is, the problem to be solved. As I said in Chap. 2, software is essentially problem-solving. In order to solve a particular problem, the programmer creates a solution, which should be universal, that is, applicable to all such problems. The programmer has not yet made categories that can be used to subsume the particular case/problem. She/he has to invent and create these categories. This universal is the set of syntactic and semantic categories (the high-level language) that the programmer uses to define and solve a particular problem. Therefore, the first work of the programmer is to "translate" the particular problem (drawing, writing, printing, etc.) into an abstract structure. This is not automated work. The programmer must choose the language that best works in relation to the problem she/he faces—for instance, building a website, a smartphone app, or a calculation program, or creating a data visualization—as well as the libraries and best data architecture. This is creative work (see Indurkhya 2017).

Now, in Kantian terms, the process that leads from the individual problem to its reinterpretation in a formal system is completely non-theoretical. In fact, this process adds nothing to the "heart" of the computer, namely, the Turing machine—the computability. As I said in the Chap. 2, the notion of computation does not exactly correspond to that of a program. Computers cannot perform any possible computation. Compared to the general set of computations, only a small subset contains the computations that can be translated into a program (a set of mathematical instructions that can be executed by a computer). Therefore, what I call the "heart" of the computer is the set of basic programming languages that are adequate to spell out the rules to perform computations, that is, the set of computations for which there are programs. In 1936, logicians Alonzo Church and Alan Turing argued that certain programming languages (Turing machine, recursive functions, and the lambda calculus) are able to spell out the instructions to perform all the computations that are computable. This set of canonical mathematical structures is the starting point of any programming activity. My thesis is that programming consists of mediating between this set of canonical mathematical structures and the concrete problem that needs to be solved in the real world. The DRJ is properly the construction of the mediation between these two poles.

I claim that using a certain algorithmic style or a high-level programming language does nothing to influence the canonical mathematical structures as well as the processor (CPU) that implements them. The CPU does not understand the high-level language, the particular problem, or human reality. The programmer reflects on the problem and proposes an algorithmic solution expressed in the high-level language. She/he provides an interpretation of the problem and its solution (the higher-order representation under which to subsume the particular). This solution has to be translated into another formal language understandable to the CPU (the binary language, 1 and 0, so-called machine language) through a compiler. Nonetheless, *the CPU does not solve that problem or implement that solution*. It only performs a series of logical operations according to the canonical mathematical structures aforementioned. It does not interact with its environment. It cannot independently solve a concrete human problem in the same way that the determining judgment cannot be used to explain the living organism. In other words, the CPU can solve the concrete human problem *only if* the latter is interpreted and translated in a certain language and rules, that is, in a certain computational architecture.[6]

Turner (2018, 67) recognizes this point:

> Without programming languages, machines would be idle devices much like cars without drivers or hairdressers without combs. [...] General problem solving in these [machine] languages is difficult and unnatural because control is limited to instructions for moving data in and out of store, and representation is performed using numbers or Boolean values. *These languages are for machines not humans.* (I emphasize)

Instead, high-level languages

> employ more abstract concepts and control features such as procedures, abstract types, functions, polymorphism, relations, objects, classes, modules, and nondeterminism. These concepts aid problem solving and enable a more natural representation of the problem domain. They operate at a distance from the physical machine, and do not depend upon the architecture of any specific machine. This is made possible by layers of translation and interpretation. (Turner 2018, 67)

This is the essential function of software: allowing the CPU to interact with the environment and solve concrete human problems. Just as the faculty of judgment exerts a function of mediation between nature and understanding, software also mediates between concrete reality and the canonical mathematical structures. This is the analogy that I propose to develop in the following pages. The art of programming is the ability to see how the Turing machine can solve a specific concrete problem. All the different parts of programming (high-level programs, compilers, interpreters, specifications, etc.) are only different ways of bringing the problem and the Turing machine as close as possible. The dual nature of software, art and mathematics at the same time, comes from this mediation task.

In other words, the programmer performs the three fundamental operations indicated by Kant in *Logik* (§6): comparison, reflection, and abstraction. The programmer compares the concrete problem and the Turing machine (let us call it its "computational resources").[7] After identifying the possible connections between the problem and the "computational resources," she/he reflects on these connections and elaborates her/his solution to the problem through the "computational resources" at her/his disposal. This solution is expressed through a formal architecture and a physical machine that implements this architecture. This solution allows us to connect the problem and the "computational resources," the concrete and the abstract. This happens in two ways: (a) *input*—software allows us to interpret the concrete problem in computational terms and in a language that can be understood by the Turing machine; (b) *output*—software allows us to interpret the physical electrical impulses produced by the CPU as (b.1) the solution of the problem and/or (b.2) the physical expressions of the Turing machine abstract operations. Software reflection allows us to think that the Turing machine "solves" that problem, even if this is not the case, because the Turing machine does not "see" that problem properly. The problem solved by the Turing machine is not "our" problem (print a paper, booking a hotel, read the newspaper, etc.) but a series of mathematical functions. Software mediation/reflection allows us to think that an abstract mathematical structure can implement physical operations.

## The Software Imagination

There are two possible objections to the analogy between software and Kantian reflective judgment that I try to draw. The first is the most immediate: "Software does not have the form of a judgment, of a proposition, *then* we cannot compare software and reflective judgment." Obviously, programming languages are not formulated in natural language and, therefore, are not based on the same structures as natural language (subject + predicate, as Kant thinks). Programming languages are formal and based on a precise syntax or grammar. However, if we consider the Kantian notion of judgment in more general terms, as a power of synthesis expressed in a certain language, then it is possible to reply to the objection. Basically, programs are acts of synthesis between different *components* in interaction, thanks to *connectors*, in order to form a *system*. Then, this formula

$$x := x + 1; y := x * y$$

is tantamount to a proposition or a set of propositions. We can translate strings of code in propositions, or a set of propositions, and vice versa, for example: "print this" or "check *x*"—imperatives.[8]

The second possible objection is that our thesis reproduces, even if in alternative terms, those of Colburn and Turner: the programmer creates the harmony between physical and symbolic, between abstract and concrete, as a sort of *deus ex machina*. This is an important objection because it gives me the opportunity to clarify a crucial point. The mediation of software does not come from a *creatio ex nihilo* in the programmer's mind.

Kant can again provide us with an important suggestion: at the roots of the three aforementioned operations (comparison, reflection, and abstraction), there is the faculty of transcendental imagination. The transcendental imagination produces the synthesis between the singular and the universal in determining judgment (transcendental schematism). It is always the transcendental imagination that produces the universal from the individual in the reflective judgment. For Kant, the transcendental imagination has a synthetic function that precedes and determines judgment and its logical forms. I do not want to analyze the Kantian doctrine of imagination, which is not the subject of this chapter (see Heidegger 1990; Sellars 1978). I want to formulate another question, which extends the comparison between software and reflective judgment that I try to formulate: How does the imagination work in software?

The programmer tackles a concrete problem and fixes requirements—what his/her goal is, what characteristics the solution should have, what the company is asking for, and so on. Her/his job is to create a formal representation of this problem and these requirements. This is an act of imagination: the programmer creates an interpretation of the problem, which can be understood by the Turing machine. Why should this act be one of imagination? It is because the Turing machine and concrete reality cannot communicate—the Turing machine cannot understand the human problem. Therefore, the programmer has to interpret the problem and create a new representation of this problem (the program) that can mediate between the problem itself (the real one) and the Turing machine. This is an act of imagination. Thus, in Kantian terms, the scope of DRJ is to reach a "free agreement" between problem, imagination (the program), and understanding (the Turing machine). This "free agreement" enables us to "see" the Turing machine as it solves that problem. The imagination enables the programmer to "see" how the Turing machine can solve the problem, even if the Turing machine cannot "see" the problem. Moreover, this agreement has to be effective. It requires implementation.

Given this, the imaginative act of software cannot be a *creatio ex nihilo*. This act must respect a series of constraints: Turing machines rules, then parameters, materials, and so forth. Furthermore, it has to (a) express a language and (b) have the ability to have a physical effect, that is, to realize a causal action on the underlying material reality (the hardware, the instrument, the surrounding environment, etc.).

Now, my hypothesis is that the imaginative act underlying software—the "seeing" the solution of the problem "in" the Turing machine—is realized through writing. Why do I choose writing? For two reasons.

First, software is writing; it is based on writing. For software, to be written is not a secondary property; it is its condition of possibility. Software would not be software if it were not written.

Second, writing already is a specific form of synthesis between the abstract and concrete. This is in two different senses:

- Writing is a synthesis between a language (abstract structures: grammar, syntax, semantics) and a material support (paper, clay, screen, etc.);
- Writing is a synthesis between language and space because it "spatializes" language, and this spatialization allows the visualization of language and, therefore, a completely new perceptive experience of the language. Spatialization and visualization allow the discovery and invention of new uses of language and new concepts.

Let us now try to better explain these two points through the concept of the "graphic mind."

## The Graphic Mind

According to Bachimont (1996, 7), writing is the name of a certain type of reason:

> [...] writing creates a spatial synopsis, allowing to identify relations and properties that remain undetectable in the linear succession of the temporality of the speech; writing shows relations which are not perceptible in orality. Indeed, by producing a spatial two-dimensionality of the content of the speech, mind can simultaneously access different parts of the content independently of the order connecting these parts in the oral flow.

Stressing the centrality of writing in knowledge, Bachimont (2000) speaks of a "graphical reason," that is, a condition of possibility of "computational reason." In doing so, Bachimont extends the results of Goody (1977), the anthropologist who most contributed to understanding the role of writing in the emergence of certain cognitive operations or ways of thinking.

What makes writing such a powerful and pervasive technology? What is the essence of writing? Goody argues that to understand the differences between cultures and peoples, one must look at intellectual technology and the means of communication, which not only allow communication but also determine its content. Goody's starting point is the same as that of Ong: language is what truly characterizes the human being. The technology that regulates the use of language and communication, therefore, has a decisive weight that must be evaluated independently of ideological considerations. Through an ethnographic and historical investigation, Goody shows that writing is the first fundamental human intellectual technology that has had enormous consequences not only on the level of knowledge but also on the political and social level. However, this, Goody points out, does not mean affirming that in societies without writing there is no original and productive intellectual activity. The nature of this activity, however, is radically different from that in societies with writing—even if Goody does not emphasize the difference between oral and written culture as radically as Ong. We can also find scientific attitudes in

African tribes, which have an entirely oral culture; it is just a different way from ours, meaning from that of humans born in a written culture.

On the basis of this theoretical approach, Goody speaks of three fundamental structures of writing: the frame or matrix, the list, and the formula.

1. The frame is a graphic construct composed of vertical columns and horizontal lines. It can have two essential functions: the organization and construction of knowledge. Two examples can be given: the family tree and the tables of truth in logic. The first is a way to organize a set of data deriving from experience (the lineage of a family). The second is instead a matrix, a mechanism that works automatically, on the basis of some rules, and which, given certain inputs, allows us to obtain outputs—the construction of knowledge. It is the visual translation of an algorithm. Other examples are the table of chemical elements or the square of oppositions in Aristotle. In another sense, the frame is the space that allows us to link a certain organization of data to a certain number of operations on them. At the same time, it is what defines data and operations, their way of being represented and measured.
2. The list is a column. Goody cites three types of lists: (a) the administrative list, which is the inventory (economic, material, managerial) or the list of certain events; (b) the list as a description of actions, that is, a series of operations and rules that develop in the future (e.g., a to-do list); and (c) the lexical list, the repertoire of letters (the alphabet), names, and words. The list has a specific physiognomy: it implies discontinuity, a finite, closed, rigid order (Goody 1977, 150–151). The list clearly breaks with the spoken language, with the fluidity and irregularity of the spoken language, and imposes a rigid graphic structure. Whoever recites a list orally must follow a fixed order and cannot make changes of any kind. The list implies the idea of hierarchy, the upper/lower dualism. It is interesting to note that the key concepts of computability theory involve the concept of a list, that is, a certain graphic structure. For example, the notion of an enumerable set is defined by Primiero (2020, 22) as the sets "whose member can be enumerated or arranged in a list." We find the same definition in

Boolos et al. (2007, 3). The list is not a text. It is not even figurative art: a drawing or a diagram, and so on. I would say that the list is *a rational design*, that is, a graphic form to represent and organize data.

3. What is the formula? It is a graphical construct that, unlike the list, is not naturally hierarchical. To the upper/lower pair, the formula replaces that identity/opposition, that is, it introduces the dimension of comparison—if the list is vertical, the formula is horizontal. Let us consider the example of the equation: the formula distinguishes two areas, two spaces, joined by the symbol "=" and imposes a certain reading direction, from left to right. Even written natural language works like this, even if it is influenced by the oral dimension. In the equation, however, the formula is completely separated from oral language and shows itself for what it really is: pure space management. The formula is not a way of representation or organization of data, but a way of expressing operations on the data.

It is interesting to note that Bachimont takes up Goody's research and extends it to the concept of computation. According to Bachimont (2010), computational reason entails an evolution of Goody's structures producing programs (lists of computations), nets (relations and communication among programs), and layers (pure forms of relations among nets, apart from individual computations). The passage from the Turing machine to the universal Turing machine exemplifies the transition from programs to nets. Similarly, the movement from the universal Turing machine to the von Neumann architecture model illustrates the movement from nets to layers. Writing is the *material a priori* that computation needs to build itself. And this a priori is *material* because it entails also a modification of matter. Writing is a mark on a material base. This act makes computational processes possible. It is a *technical a priori*.

How can writing produce its conceptual structures? In a Kantian way, Bachimont thinks that writing is a space-time synthesis—he says: *une synthèse synoptique de l'écriture*. Writing is a synthesis between permanence in time and space. It synthesizes, on the one hand, memory (the need for unity and order in the temporal flow) and, on the other, a kind of organization based on simultaneity. In writing we have a *spatialisation of time*. The temporal development of discourse (linguistic acts in general) is translated into a spatial organization: in lists, tables, and formulas, the items are defined by their position in an abstract space. Writing gives us a synthetical way to grasp together dispersed items in the temporal flow. Lists, tables, and formulas are the main graphic schemes by which writing can realize this space-time synthesis. In this way, "writing can show relations that cannot be perceived by listening to a voice" (Bachimont 1996, 15; translation is mine). The same is the case for computation. These graphical schemes mark and modify matter. From the point of view of technology, this is their *telos*.

Therefore, I distinguish six structures (types of spatial synopsis) to which writing gives rise. I summarize them in the following table. The structures of computational reason are derivative of those of graphical reason. The first ones were described by Goody, while the second by Bachimont (I changed his vocabulary a bit). These six structures are the essence of writing. They are not abstract concepts, nor material tools (such as knives, washing machines, or other). They are material and immaterial at the same time. Materiality allows thought (and linguistic structures in particular) to apply to experience. Immateriality allows thought to be modified by experience (Table 5.1).

While in the graphical reason the three structures (list, formula, and table) can work separately, in computational reason, no: they must interact. Therefore, following Bachimont's indication, I claim that software arises from the union of the three structures of computational reason. This union makes the programming language possible. In Kantian terms, stack, string of code, and net are three syntheses of language and space that make possible unifying a concrete problem and a computational language, that is, one complying Turing machine.

However, there is another aspect that we must underline in order to understand the unity of graphic reason (Goody) and computational reason (Bachimont), namely their visual character. Writing is essentially a visual gesture.

**Table 5.1** The fundamental structures of graphic reason and computational reason

| Graphical reason | List | Formula | Table |
|---|---|---|---|
| Computational reason | Stack | String of code | Net |

Graphics is the processing of information through the visual medium; "Graphic representation is the transcription, in the graphic system of signs, of information known through the intermediation of any system of signs" (Bertin 1983, 8; my translation). For this reason, the graphic representation implies an analysis of the information and the distinction, in it, between variable and invariable elements. Its three key functions are (a) to record information, then to create an artificial memory that supports the effort of memorization; (b) to communicate information, that is, the actual storage and transmission of information; (c) to treat information, then simplify the information and translate it into a 2- or 3-dimensional image (Bertin 1983, 13–14). Bertin shows that it is possible, starting from the three basic elements of the image (point, line, area), to construct a grammar of graphic representation, distinguishing the means, the rules, and the conditions of readability. He also distinguishes three large groups of graphical representations: diagrams, networks (organization charts, trees, classifications, etc.), and maps (the main example is the geographical map). Writing is at the same time an element of graphic representation and a graphic representation itself.

However, in reality, Bertin speaks very little of writing as a graphic representation. My thesis is that it is possible to show how the structures of writing derive from Bertin's rules of visual grammar.

The link between visual grammar and writing is the concept of diagram, which plays a crucial role in our strategy. My claim is that writing is a set of diagrams. This set of diagrams allows the reorganization of experience and language. But what is a diagram? Diagrams are necessary to rational thought. As Krämer says, "diagrams and related graphical artifacts—like scripts, lists, tables, graphs, and maps—lay the foundation for acquiring and evaluating knowledge, a foundation which is indispensible to scientific inquiry" (Krämer and Ljungberg 2016, 163).

According to Stjernfelt (2007), the diagram is an image connected to reasoning; it is an image whose continuous transformation allows us to discover new aspects of reality—new information. Stjernfelt (2007) compares Peirce's notion of diagram to Husserl's concept of eidetic intuition. The diagram allows, thanks to its continuous transformation, to grasp the essential structures of the experience. Stjernfelt claims that "as soon as an icon is contemplated as a whole consisting of interrelated parts whose relations are subject to experimental change, we are operating on a diagram" (Stjernfelt 2007, 92).

I am more interested in Stjernfelt's general definition: the diagram is an image whose continuous variation allows us to discover new things. I think that this fits very well with writing. Writing is a set of diagrams, that is, letters of the alphabet that continuously vary. The combinations of letters continuously vary in order to adapt to the different situations and get new information. This continuous variation gives writing a heuristic power—even if today we hardly notice it anymore. Writing is diagrammatic reasoning. Writing is the means of a logic of discovery, and this is especially true for mathematics. As Peirce says (cited in Stjernfelt 2007, xiii):

> The first things I found out were that all mathematical reasoning is diagrammatic and that all necessary reasoning is mathematical reasoning, no matter how simple it may be. By diagrammatic reasoning, I mean reasoning which constructs a diagram according to a precept expressed in general terms, performs experiments upon this diagram, notes their results, assures itself that similar experiments performed upon any diagram constructed according to the same precept would have the same results, and expresses this in general terms. This was a discovery of no little importance, showing, as it does, that all knowledge without exception comes from observation.

Peirce emphasizes more the cognitive and heuristic power of the diagram. We can interpret this point in two ways. In a broad sense, a map of the London Underground provides us with knowledge: by reading it, we understand what way to take it and how to get to its destination. The diagram speaks to us of an object and lets us know certain properties and their internal relations. In another, more restricted sense, the diagram is a tool of knowledge, capable of producing new knowledge and reasoning: this is the function of the diagram in constructive geometrical proof or in the research and teaching of mathematics. In this second case, we get to know something more about the object, thanks to the possibility of continual transformation of the diagram. Diagrams are kinesthetic tools open to change and manipulation.

Peirce's diagram has a double nature: perceptive and general, type and token, abstract and manual. First of all, the diagram is an icon, that is, a sign referring to its object by virtue of similarity. However, the diagram is not just an icon, like other icons such as images or metaphors. Indeed, the diagram presents two main characteristics: (a) *the relationality*, the similarity with the object is due to the fact that the diagram is a skeleton of the relationships between the parts of the object represented; (b) *the abstraction*, the diagram arises from an effort of abstraction with respect to the real object. According to Pierce, we can find a diagram in every image. We have to just consider the relations between the parts of the object as a whole and isolate them. We will obtain a skeleton that is particular and general at the same time. For instance, a triangle that I draw to make a geometric demonstration is at the same time a particular, individual drawing and the representation of the triangle in general—that type of triangle. Now, transforming that triangle (changing a part of it or rotating it, etc.) allows us to better understand the nature of the triangle, demonstrate its properties, and discover something new. This is the function of the diagram. There is ambivalence, "even a struggle, between reasoning and imagination in Peirce's attempt at disregarding the materiality and visibility of diagrams since, for Peirce, rational thought should steer the viewer's imagination. And yet, Peirce cannot elide the function of experience and imagination in his concept of diagrammatic reasoning" (Krämer and Ljungberg 2016, 34).

Let us better analyze Peirce's formulation:

> A diagram is an icon of a set of rationally related objects. By rationally related, I mean that there is between them, not merely one of those relations which we know by experience, but know not how to comprehend, but one of those relations which anybody who reasons at all must have an inward acquaintance with. This is not a sufficient definition, but just now I will go no further, except that I will say that the Diagram not only represents the related correlates, but also, and much more definitely represents the relations between them, as so many objects of the Icon. Now necessary reasoning makes its conclusion evident. What is this 'Evidence'? It consists in the fact that the truth of the conclusion is perceived, in all its generality, and in the generality of the how and the why of the truth is perceived. [...]

> the Diagram remains in the field of perception and imagination; and so the Iconic Diagram and its Initial Symbolic Interpretant taken together constitute what we shall not too much wrench Kant's term in calling a Schema, which is on the one side an object capable of being observed while on the other side it is General. (Of course, I always use 'general' in the usual sense of general as to its object. If I wish to say that a sign is general as to its matter, I call it a Type, or Typical.) (Stjernfelt 2007, 23)

The diagram is a cognitive tool. Its organization and constant variation allow information to be creatively reorganized, or to create new ones. The concept of diagram implies phenomenological realism: access to reality is not a problem because the subject is already in reality, part of it (the concept of intentionality). The problem, therefore, is not whether the subject accesses reality, but how it represents and organizes the data that come from reality, as well as how to discover new things from them. However, the representation and organization of data are expressed not in abstract structures, but through concrete tools, material drawings.

> For semiotics, this central role of the sign type of diagrams implies a basic realism. It is indeed possible to acquire and develop knowledge about different subjects by the construction and manipulation of diagrams charting those subjects—based on the fact that the structure of these diagrams are, in some respects, similar to the structure of their objects. This similarity does not have to be evident for a first glance [...] The recognition of the foundational role of diagrams in thinking thus immediately implies the recognition of the iconicity in thought and signs—the similarity between sign and object which is especially developed and constrained in the diagram case. (Stjernfelt 2007, xiii)

Now, without the application of visual-imaginative grammar (point, line, area), diagrams and diagrammatic reasoning could not exist. The structures of the graphic mind derive from the application of visual grammar to the process of building diagrams.

We can, at this point, construct a complete "graphic transcendental deduction" of the computation (Fig. 5.1).

I claim that each type of programming language comes from the interaction between a language (a syntax, an alphabet, and a set of rules) and

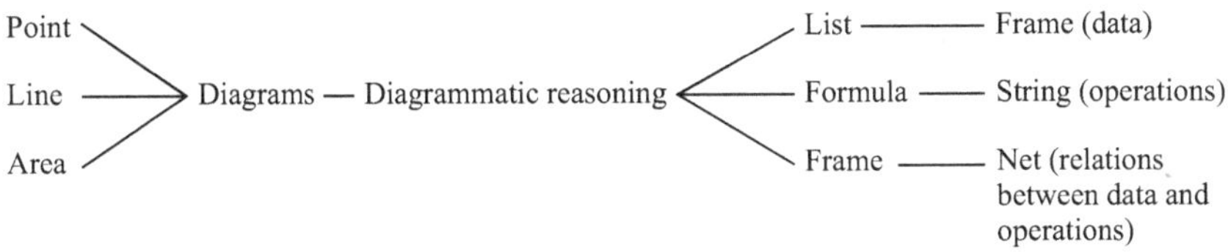


**Fig. 5.1** The structures of the graphic mind derive from the application of the basic visual grammar to the process of building diagrams

the three conceptual structures of "computational reason." The stack is the basic form of data architecture: it allows the classification and spatial organization of data. The net allows communication between different stacks and combines stacks in a coherent whole. The string of code activates the relationships between the objects in the stacks. For instance, in Java programming language, the stacks correspond to classes, which include attributes and methods. Each string of code combines a method and an object and, therefore, makes the stacks interact through the net. The same thing can be said for other types of programming languages. My claim is that any syntax and semantics of software language presuppose these conceptual structures. When the programmer uses terms such as "object" or "operation," she/he constructs their meaning through these syntheses of language and space. She/he just expanding her/his diagrammatic reasoning.

The programmer could not draw thi

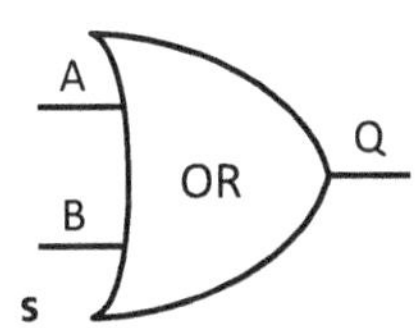


or write this

```
Car
make:String,  model:String,  engine:Num
Find Speed,  Start,  Stop.
```

if she/he did not previously have the structures of diagrammatical and graphical reason and, then, those of computational reason. The art of programming is first an expression of diagrammatical and graphical reason.

As I mentioned before, the structures of computational reason are placed between the concrete problem to solve and the Turing machine. In other words, they mediate the relationship between "understanding" in Kantian terms (the Turing machine) and "life" (the concrete problem to solve, the implementations, etc.). The functional and physical levels communicate, thanks to the mediation of these spatial syntheses. Moreover, thanks to its physicality, software can causally act on the physical machine (circuits) and have an effect in the world. I am not saying that the synthesis of the functional and physical levels *is* writing, only that *it is realized through* the spatial syntheses made possible by writing. It is thanks to the hybrid (conceptual and material) nature of these three spatial syntheses that we can "see" the functional level in the physical, the series of mathematical operations in the electrical impulses produced by the CPU, and, therefore, the Turing machine solving that problem.

## Writing Is Technology: An Incursion in the Actor-Network Theory

The actor-network theory (ANT) is a theoretical model that intends to understand science and technology together, or rather *technoscience*. It is also a general social theory centered on technoscience and also, especially in Latour, a philosophical account of the world.

Writing plays a crucial role in Latour's ANT—at least, in some of his most important works. It is useless to try to give a complete view of Latour's work, which is multifaceted and multicentric; impossible to choose a unique label. However, there is an underlying inspiration that cannot be overlooked: Latour claims that several of the established conceptual distinctions used to define the modern world—for example, nature versus society, and facts versus values—provide at best little guidance to understanding what goes on in science, law, politics, and religion and more likely will lead us astray. Latour therefore intends to provide a re-description of these concepts through an ethnographic work and a deep materialist philosophy. Latour reveals the complexity of the scientific fact by "opening it," that is, by showing the network of mediations and actors, human and otherwise, that make it possible. Therefore, ANT represents technoscience as the creation of larger and stronger networks of actors interacting with each other. "Just as a political actor assembles alliances that allow him or her to maintain power, so do scientists and engineers"; however, "the actors of ANT are heterogeneous in that they include both humans and non-humans entities, with no methodologically significant distinction between them" (Sismondo 2014, 289).

In *Laboratory Life*, co-written with Steve Woolgar, Latour shows that writing, understood in its original sense of trace and recording, is an essential element in the construction of scientific facts. The drafting of documents, starting from other documents and other traces, is the central activity of scientists in the laboratory and crosses and conditions all the others. Latour shows that in the Salk Institute—where the investigation behind the book was carried out—writing is present everywhere; even the humblest technicians in their work on animals or chemicals must compulsively and obsessively write, take notes, and record facts. Scientists and technicians write numbers, codes, labels, fill invoices, collect data, and compose reports and cards, up to drafting abstracts and papers to be published, or projects to request funding. The laboratory is a written world. A chemical substance becomes a scientific object and therefore a fact when it is translated into a process and a series of traces: it is cataloged, encrypted, illustrated with a diagram or with a slide, inserted into an argument as evidence, which, if approved, will attain the final form of a paper to be published in a journal. The paper will allow other writing, other papers, or annotations, and will thus expand that great written object that is the library.

Latour argues that the scientific fact is first and foremost, but not only, a literary construction, which must therefore be treated with the methods of semiotics and linguistics. In this construction, writing is transformed by passing through a series of technical mediations, debates, and comparisons between colleagues. Writing, a technology, functions as an organizational principle that gives meaning to observations and connects them to practices. In *Laboratory Life*, Latour meticulously describes different types of writing: by hand, by machine, printed, through the use of the computer, but also translated in terms of visualization (curves, graphs, etc.). Photography is also considered a writing, a form of trace. He talks of "inscription device[s]" (51) that "transform pieces of matter into written documents" (51). More exactly, "an inscription device is any item of apparatus or particular configuration of such items which can transform a material substance into a figure or diagram which is directly usable by one of the members of the office space" (51). At the origin of the scientific fact and of reality, for Latour, there is, therefore, writing performed by machines. The "inscription devices" are capable of translating any material reality—the object of the research—into a series of recordable traces; for example, the recording of a sound or the measurement of the slope of a terrain are performed through machines that translate that phenomenon into traces. On this basis, many other forms of writing are used to define the final product: a paper, a report, a poster for a conference or a project, which will, in turn, be measured through another writing, that is, the quotation. This is the central point of Latour's methodological approach: a non-human, a hybrid, that is, writing, is the starting point for building the relationship between the subject and the object.

What exactly happens in this long process of writing and re-writing? According to Latour, a work of producing and transforming statements takes place. The statements resemble particles that move according to a Brownian motion: the equilibrium represents the scientific fact. When equilibrium has been reached and the fact settled, it is no longer questioned. It is a black box, which can no longer be opened. The discussion stops. To reopen the box requires restarting the discussion and therefore questioning all the work done. Through writing, therefore, scientists fight a war of position whose goal is to persuade others (competitors) of the correctness of their statements. The activity of the laboratory is "the organization of persuasion through literary inscription" (88). Each scientist tries to improve his position by getting published and thus obtaining a better job.

To better understand this idea, we need to analyze two key concepts of Latourian anthropology. The first is what I will call the priority of the dispute. This idea can be summarized from the Latourian *summa,*

*Reassembling the Social*: "if agencies are innumerable, controversies about agency have a nice way of ordering themselves" (Latour and Woolgar 2005, 52). The sociologist must begin his work by analyzing the controversies between the actors (human and otherwise) in the creation of groups and in the achievement of relevance in their own groups. For this reason, Latour brings politics into science and science into politics: this is evident in the work on Pasteur (Latour 2011). In other terms, ANT focuses on agency and power, and so it "may encourage the following of heroes and would-be heroes" (Sismondo 2014, 294). This consideration leads us to another important aspect: ANT remains essentially structuralist—an object is defined by its position in the network—albeit in a very different sense from classic Saussure-inspired structuralism. In fact, the paradigm of difference is replaced by that of translation and the alliance. It is not society that explains technology (and more), but the opposite: it is technology that explains society precisely, thanks to its power to create associations between actors and translate the interests of the actors in order to create consensus. This type of association founds society.

The second aspect that I want to emphasize is what some have called "flat ontology," that is, an ontology that puts all entities (humans and non-humans) on the same footing without reducing one to the other; we can never explain an entity as the result of another, as the effect of something else. "Latour's guiding maxim is to grant dignity even to the least grain of reality" (Harman 2009, 15). It is a flat ontology in the sense that it does not impose any classification or hierarchy from above; it is completely symmetrical. This is the idea of "democracy extended to things" (Latour and Woolgar 2007, 22). The world is an immense network defined by actors who move in it and each has its own dignity: "Atoms and molecules are actants, as are children, raindrops, bullet trains, politicians, and numerals" (Harman 2009, 14). One might think that, in this way, Latour intends to re-propose the Aristotelian principle according to which entities are individuals—that is, a nucleus (the substance) that resists change and the attributes that connect faces. This is not the case. For Latour, entities are not entities but sets of strength and resistance in constant interaction with others; "actants are not stronger or weaker by virtue of some inherent strength or weakness harbored all along in their private essence. Instead, actants gain in strength only through their *alliances*" (Harman 2009, 15). For Latour, "an object is neither a substance nor an essence, but an actor trying to adjust or inflict its forces, not unlike Nietzsche's cosmic vision of the will to power" (ibid.). From this point of view, there is no dualistic object-subject relationship, fixed and absolute. Subject and object, as well as their own relationship, are socially constructed, in the sense that they are immense nets of power relations; understanding a phenomenon, an entity, means making evident these relations of force that involve actors of all kinds: atoms, bodies, political forces, technical objects, and so on.

From the ANT perspective, writing is something absolutely unique. In fact, writing is not a technology like any other. Writing is hybrid: neither subject nor object, neither human nor non-human, neither private nor social, neither abstract nor concrete. Writing is at the same time an actor like any other and what allows the power relations between the actors and is therefore the basis of society. As Latour says, society "is visible only by the *traces* it leaves (under trials) when a new association is being produced between elements which themselves are in no way 'social'" (2005, 15). In *L'espoir de Pandore*, Latour states that there is no society without technique because technology stabilizes and regulates social relations. "The techniques do not presuppose the existence of society, but only of a semi-social organization that puts non-humans from very different times, places and materials in contact" (2007, 221–222). Now, writing is the condition of the technique, not the other way around. There is technique because there is writing, trace, inscription. We could almost say that writing is the actor *par excellence*. This is also demonstrated by the analysis that Latour and Woolgar (1979) make of Pasteur's text, which is an exercise in semiotics applied to a non-fiction text.

## The Software Delight

In the first book of the *Critique of Judgement* (§1), Kant (2016, 65) writes,

> If we wish to discern whether anything is beautiful or not, we do not refer the representation of it to the object by means of understanding with a view to cognition, but by means of the imagination (acting perhaps in

> conjunction with understanding) we refer the representation to the subject and its feeling of pleasure or displeasure. The judgement of taste, therefore, is not a cognitive judgement, and so not logical, but is aesthetic, which means that it is one whose determining ground cannot be other than subjective.

The judgment of taste is based on a certain agreement between imagination and understanding. This means that the judgment of taste is based on the subject's feelings, that is, the manner in which the subject is affected by representations. The judgment of taste applies the principle of finality to the delight coming from the agreement between imagination and understanding, which Kant calls a "free play" of faculties. This delight is independent of any interest (see §2). In this delight caused by representations, the faculty of judgment finds the mark of finality. As Kant (2016, 67) says,

> This relation, present when an object is characterized as beautiful, is coupled with the feeling of pleasure. This delight is by the judgement of taste pronounced valid for everyone; hence an agreeableness attending the representation is just as incapable of containing the determining ground of the judgement as the representation of the perfection of the object or the concept of the good. We are thus left with the subjective finality in the representation of an object, exclusive of any end (objective or subjective) consequently the bare form of finality in the representation whereby an object is given to us, so far as we are conscious of it as that which is alone capable of constituting the delight which, apart from any concept, we estimate as universally communicable, and so of forming the determining ground of the judgement of taste.

In DRJ, the agreement between imagination and understanding, which is realized through writing, delights the programmer. The programmer enjoys when the machine runs well and fast and solves the problem. This delight is seen as the expression of a finality: the machine acts for us and improves our world. Nevertheless, this finality is purely subjective. The machine is built by humans and responds to human purposes. Software specifications meet human criteria.

I see here an interesting parallelism between DRJ and the Kantian judgment of taste. In both cases, the agreement between imagination and understanding generates a delight that is the expression of the principle of finality. As we said earlier, the fundamental task of a programmer is to translate a problem (the singular concrete case) into computational terms (the understanding). She/he can do this only by using her/his imagination because there is no connection between concrete reality and the Turing machine. The programmer has to use her/his imagination. She/he creates the universal by which to think about the single individual case and make it comprehensible through understanding (the Turing machine). In doing so, she/he uses the imaginative structure of writing, the form of spatialization, and the materialization of language. The programmer reaches her/his scope only when she/he reaches an agreement between the concrete problem, her/his imagination, and the Turing machine. The program mirrors this agreement, which in turn makes the physical machine work and produces a delight. Through this delight, DRJ applies the principle of finality.

In the judgment of taste, the principle of finality is not the result of the operation of judging; it is its condition of possibility, namely, what guides the power of judging and makes it applicable. The same can be said for software. In her/his imaginative work, the programmer is oriented and guided by the principle of finality. This principle precedes and determines the syntheses of writing between the physical and functional levels. It precedes and determines all the stages of the programming.

The main way in which the principle of finality appears in programming is the act of design. Programming is essentially a design act: "design is everywhere in computer science" (Turner 2018, 128). The choice of what language to use and the algorithmic style is strictly connected to design choices and is, therefore, aimed at the construction of well-designed programs. Design is not just about beauty. "Design is a practice of creation turned towards the future and supported by an innovative intention" (Vial 2010, 44).

According to Turner (2018, 161), the hallmarks of a good digital design are (1) simplicity, (2) expressive power, and (3) security. However, Turner subordinates design to semantics: "More explicitly, the things that we may refer to and manipulate, and the processes we may call upon to control them, need to be settled before any actual syntax is defined. This is the 'semantics-first principle,' according to which, one does not design a language, and then proceed to its semantic definition as a post hoc endeavor; semantics must guide design" (169).

Even on this point, Turner's analysis appears to be characterized by an excessive intellectualization. As the French philosopher Stéphane Vial (2010) suggests, the objects of design are objects that have been submitted to a process of design, which consists of conceiving and producing effects that point to "experiences to be lived by means of forms" (115). The effects of design operate on the level of form, social meaning, and experience; thus, Vial sees design primarily as a "generator of human existence that proposes possible experiences" (65). In Vial's view, design deals not so much with the being as with events, not so much with the existing as with the new that will emerge. As the French designer Alain Findeli (2010) writes, the purpose of design is to improve the *habitabilité du monde*, that is, our ability to live on this planet. Thus, design has a phenomenological and existential function, the aim of which is to improve the interaction between machines and humans and, therefore, between humans and the world. In this sense, design can be considered an extension of the use of Kant's principle of finality in the judgment of taste. The agreement between imagination and understanding produces a delight that is the expression of a finality in nature. Programmers' work makes the interaction between humans and machines possible through design as an expression of the principle of finality. Design is "a means for human beings to envision and realize new possibilities of creating meaning and experience and for giving shape and structure to the world through material forms and immaterial effects" (Folkmann 2013, 45).

From this point of view, I claim that design is the condition of programs, not the opposite. The programmer does not decide abstractly which objects to take into consideration: she/he deals with problems and has to choose the best strategy to solve them and give them meaning. The design criteria of elegance, correctness, simplicity, uniformity, modularity (the process of breaking up complex problems into smaller, simpler ones.), transparency, reliability, and so forth shape the semantic and syntactic of the program. All possible criteria of the correctness of software are thought by the programmer through the principle of finality, that is, through design. Writing is the first means by which software design is achieved.

I think that this view is closer to the way in which programmers understand their work. "One of the main reasons most computer software is so abysmal is that it's not designed at all, but merely engineered. Another reason is that implementors often place more emphasis on a program's internal construction than on its external design, despite the fact that as much as 75 per cent of the code in a modern program deals with the interface to the user" (Kapor 1996, 5). Moreover, software is not just a design job. It is also the source of a new form of design. "A discipline of software design must train its practitioners to be skilled observers of the domain of actions in which a particular community of people engage, so that the designers can produce software that assists people in performing those actions more effectively" (Denning and Dargan 1996, 112).

As Fred Brooks (1975, 25) claims

> Why is programming fun? What delights may its practitioner expect as his reward? First is the sheer joy of making things. As the child delights in his mud pie, so the adult enjoys building things, especially things of his own design. I think this delight must be an image of God's delight in making things, a delight shown in the distinctness and newness of each leaf and each snowflake. Second is the pleasure of making things that are useful to other people. Deep within, we want others to use our work and to find it helpful. In this respect the programming system is not essentially different from the child's first clay pencil holder "for Daddy's office." Third is the fascination of fashioning complex puzzle-like objects of interlocking moving parts and watching them work in subtle cycles, playing out the consequences of principles built in from the beginning. The programmed computer has all the fascination of the pinball machine or the jukebox mechanism, carried to the ultimate. Fourth is the joy of always learning, which springs from the nonrepeating nature of the task. In one way or another the problem is ever new, and its solver learns something: sometimes practical, sometimes theoretical, and sometimes both.

## Conclusions

The purpose of this chapter was to strengthen the philosophical foundation of the theses developed in the first two chapters of this book.

In this chapter, I proposed a definition of software from a transcendental Kantian perspective, that is, through the concept of reflective judgment. I explain why and how we can consider software as a new form of reflective judgment, "digital reflective judgment." This judgment is realized through a type of imaginative synthesis that mediates between physical implementations and mathematical structures. I identified these structures as specific forms of writing that I called "graphical and computational reasons," following Goody (1977) and Bachimont (2000) . Finally, I clarified my approach by showing the parallelism between software and the Kantian judgment of taste. In both cases, the principle of finality is an a priori condition.

I think that a Kantian approach to the question of software is a good model in explaining the nature of software in accordance with the concrete work of programmers. The transcendental approach to software avoids the main difficulties of Turner's approach outlined in section "The Kantian Reflective Judgment." In fact, I have shown that a transcendental approach is able to (1) explain the interaction between software and users through design; (2) escape the overly static framework of an ontology of the thing and then support an ontology of the process, which is much more suitable in explaining a phenomenon such as software; and (3) avoid an excessive intellectualization of software and highlight its creativity and the underlying work of the imagination.

## Notes

1. I quote using the relevant volume and page number from the standard "Akademie" edition of Kant's works: *Kants Gesammelte Schriften*, edited by the Königlich Preussischen (now Deutschen) Akademie der Wissenschaften (Berlin: G. Reimer [now de Gruyter], 1902–).
2. For the meaning of these expressions in Kant, see Eisler (1994, 48–54). For Kant, there are three types of judgment: analytical a priori, synthetic

a posteriori, and synthetic a priori (see Eisler 1994, 585ss). The supreme principle of all synthetic judgments is that "every object is subject to the necessary conditions of the synthetic unity of the different intuitions in a possible experience. [...] The conditions of the possibility of experience in general are at the same time conditions of the possibility of the objects of experience, and for this they have an objective validity in a synthetic judgment a priori" (AK III 39–40); "[a]ll analytical judgments rest entirely on the principle of contradiction and are by nature a priori knowledge [...]" (AK IV, 266–267).

3. I am aware that this description is schematic and does not show the real complexity of Kant's thought on this topic. However, what interests me in this section is to highlight only the fundamental points of Kant's theory of judgment and then focus on the distinction between determinant and reflective judgment.
4. One must not make the mistake of thinking that the *Critique of Pure Reason* deals only with determining judgment and, instead, that the *Critique of the Judgment* deals only with reflective judgment as if, in Kant, there was a clear distinction between these two uses of propositional contents. It must be emphasized that Kant, in the third *Critique*, defines aesthetic and teleological judgments as *only* reflexive in the sense that these judgments are *entirely* reflective. Many other judgments are determining and reflective at the same time (Longuenesse 1993, 208–215).
5. The problem of analogy in Kant is very complex. I do not want to tackle this issue here. In Kant, there are several ways in which the term "analogy" is used. I would say that we can distinguish three main meanings: theological, cognitive (the analogies of experience), and mathematical. See Callanan (2008).
6. Computable functions are precisely those computable by lambda terms or general recursive functions. Alonzo Church and Alan Turing published independent papers that purported to demonstrate a general solution to the *Entscheidungsproblem*. A good number of solutions were proposed that all turned out to be extensionally equivalent. Obviously, this is not the place to deal comprehensively with computability: I refer mainly to Turing (1936), Adams (1983), Copeland et al. (2013), Immerman (2011), Boolos et al. (2007). Regarding the analogy proposed in this chapter, I consider computability as an abstract mathematical structure and software as a way of representing and interpreting this structure. I compare

computability to the Kantian understanding: it is a set of mechanical laws that govern our way of thinking about the world.

7. I use the expression "computational resources" here because there are many ways to understand computation and many ways to implement it.
8. The architecture of programs is determined, above all, by the paradigm that the programmer decides to follow. It can be *imperative* (or *procedural*), *functional*, *logic*, or *object-oriented*, each of which is connected to a precise conception of the program and of computation. Nevertheless, many languages can be mixed (Turner 2018, 67–76).

# References


Adams, R. 1983. *An Early History of Recursive Functions and Computability. From Gödel to Turing*. Boston: Docent Press.

Bachimont, B. 1996. *Signes formels at computation numérique*. http://www.utc.fr/~bachimon/Publications_attachments/Bachimont.pdf.

———. 2000. Engagement sémantique et engagement ontologique: conception et réalisation d'ontologies en ingénierie des connaissances. Ingénierie des connaissances: évolutions récentes et nouveaux défis 4: 305–323.

———. 2010. *Le sens de la technique: le numérique et le calcul*. Paris: Encre Marine.

Bertin, J. 1983. *Semiology of Graphics*. Madison, WI: University of Wisconsin Press.

Boolos, G., J. Burgess, and Richard C. Jeffrey. 2007. *Computability and Logic*. Cambridge: Cambridge University Press (1st ed. 1974).

Brooks, F. 1975. *Mythical Man-Month*. Boston: Addison-Wesley.

Callanan, J. 2008. Kant on Analogy. *British Journal for the History of Philosophy* 16 (4): 747–772.

Copeland, Jack B., Carl J. Posy, and Oron Shagrir, eds. 2013. *Computability. Turing, Gödel Church, and Beyond*. London; Cambridge, MA: MIT Press.

Denning, P., and P. Dargan. 1996. Action-centered Design. In *Bringing Design to Software*, ed. T. Winograd, 105–120. New York: ACM Press.

Eisler, R. 1994. *Kant Lexicon*. Translated by A.-D. Balmès and P. Osmo. Paris: Gallimard.

Findeli, A. 2010. Searching for Design Research Questions: Some Conceptual Clarifications. In *Questions, Hypotheses & Conjectures: Discussions on Projects by Early Stage and Senior Design Researchers*, ed. R. Chow, W. Jonas, and G. Joost, 23–36. London: Bloomington.

Folkmann, M.N. 2013. *The Aesthetics of Imagination in Design*. Cambridge, MA: MIT Press.

Ginsborg, H. 2006. Empirical Concepts and the Content of Experience. *European Journal of Philosophy* 14: 349–372.

Goody, J. 1977. *The Domestication of the Savage Mind*. Cambridge: Cambridge University Press.

Hanna, R. 2001. *Kant and the Foundations of Analytic Philosophy*. Oxford: Clarendon/Oxford University Press.

———. 2005. Kant and Nonconceptual Content. *European Journal of Philosophy* 13: 247–290.

———. 2006. *Rationality and Logic*. Cambridge, MA: MIT Press.

———. 2017. Kant's Theory of Judgement. In *The Stanford Encyclopedia of Philosophy*, ed. E. Zalta. https://plato.stanford.edu/entries/kant-judgment/supplement4.html.

Harman, G. 2009. *Prince of Networks: Bruno Latour and Metaphysics*. re.press.

Heidegger, M. 1990. *Kant and the Problem of Metaphysics*. Translated by R. Taft. Bloomington: Indiana University Press.

Immerman, Neil. 2011. Computability and Complexity. In *The Stanford Encyclopedia of Philosophy*, ed. E. Zalta. https://plato.stanford.edu/entries/computability/.

Indurkhya, B. 2017. *Some Philosophical Observations on the Nature of Software and Their Implications for Requirement Engineering*. https://www.academia.edu/7817075/.

Kant, I. 2016. *The Critique of Judgement*. Translated by J. Creed Meredith. Scotts Valley: CreateSpace.

Kapor, M. 1996. A Software Design Manifesto. In *Bringing Design to Software*, ed. T. Winograd, 1–9. New York: ACM Press.

Krämer, S., and C. Ljungberg. 2016. *Thinking with Diagrams*. Boston; Berlin: De Gruyter.

Land, T. 2011. Kantian Conceptualism. In *Rethinking Epistemology*, ed. G. Abel et al., 197–239. Berlin: DeGruyter.

———. 2015. Nonconceptualist Readings of Kant and the Transcendental Deduction. *Kantian Review* 20: 25–51.

———. 2016. Moderate Conceptualism and Spatial Representation. In *Kantian Non-conceptualism*, ed. D. Schulting, 145–170. London: Palgrave Macmillan.

Latour, B., and S. Woolgar. 2011. (1984; Engl. trans. 1986) Pasteur: guerre et paix des microbes, suivi de Irréductions. Paris: La Découverte.

———. 1979. *Laboratory Life. The Social Construction of Scientific Facts*. Los Angeles: Sage.
———. 2005. *Reassembling the Social*. Oxford: Oxford University Press.
———. (1999) 2007. L'espoir de Pandore. Paris: La Découverte.
Longuenesse, B. 1993. *Kant et le pouvoir de juger*. Paris: Puf.
Mahoney, M. 2005. The Histories of Computing. *Interdisciplinary Science Reviews* 30 (2): 119–135.
Peirce, C. S. 1992. *The Essential Peirce*, vol I. (1867–1893). Edited by N. Houser and C. Kloesel. Bloomington: Indiana University Press.
———. 1998. *The Essential Peirce*, vol II. (1893–1913). Edited by N. Houser and C. Kloesel. Bloomington: Indiana University Press.
Possati, L.M. 2020. Digital Reflective Judgement. A Kantian Perspective on Software. *Critical Hermeneutics* 4 (1): 1–34.
Primiero, G. 2020. *On the Foundations of Computing*. Oxford University Press.
Sellars, W. 1978. The Role of the Imagination in Kant's Theory of Experience. In *Categories: A Colloquium*, ed. H.W. Johnstone Jr., 120–144. Pennsylvania State University.
Sismondo, S. 2014. Actor-Network Theory: Critical Considerations. In *Philosophy of Technology. The Technological Condition*, ed. R.C. Scharf and V. Dusek, 289–296. Oxford: Wiley.
Stjernfelt, F. 2007. *Diagrammatology*. Berlin: Springer.
Turing, A.M. 1936. On Computable Numbers, with an Application to the *Entscheidungsproblem*. *Proceedings of the London Mathematical Society* 42: 230–265.
Turner, R. 2018. *Computational Artifacts. Towards a Philosophy of Computer Science*. Berlin: Springer.
Vial, S. 2010. *Court traité du design*. Paris: Puf.

# 6

# Conclusions: Toward a Critical Code Thinking

## Software: Its Nature and History

This book has been about software, its nature and history. The crucial belief behind each part of the book is that software is not only a philosophical object worthy of in-depth analysis but also the condition of possibility of the digital world. If we want to understand the digital world, we must first understand software.

This book is halfway between two expanding fields of study: digital hermeneutics (Romele 2019) and critical code studies (Marino 2020). Software hermeneutics is part of the digital hermeneutics research program, and this is because "digital hermeneutics is a part of material hermeneutics dealing with a specific kind of technology, namely digital technologies" (Possati and Romele 2021). Digital hermeneutics can be seen "as a series of considerations on the most immediate and empirical aspects of digital media and technologies, in particular, but not exclusively, in what concerns their ways of mediating between the humans and the world" (Possati and Romele 2021). Digital technologies have hermeneutic effects, that is, they give new interpretations of our way of being in the world. However, these technologies are hermeneutics in themselves because they presuppose worldviews and patterns of meaning. Critical



L. M. Possati, *Software as Hermeneutics*,
https://doi.org/10.1007/978-3-030-63610-4_6

code studies share this thesis with digital hermeneutics, even if they focus on the concrete study of the code and its applications. However, critical code studies demonstrate that the main theses of digital hermeneutics are valid. Reading the code is an operation that requires specific methodologies. The interpreter must analyze the comments of the programmers interpolated in the source code, as well as the metaphors and rhetorical forms.

Software hermeneutics proposes an interaction between these two fields: the critical analysis of the code from a philosophical, ontological, and ethical perspective.

The starting point of this book was the fracture between living experience and the code. In the first chapter, I highlighted this aspect from a phenomenological point of view. Software is the origin of the digital experience, but it remains hidden, invisible. The analysis of this fracture led me to question the nature of software as such, and therefore the possibility of a software ontology. For this reason, in the first chapter, I formulated a first definition of software by distinguishing three essential aspects: algorithm, problem, and code. However, this definition was incomplete because it did not explain the unity of these three aspects and, above all, the synthesis between software and lived experience. At the end of the first chapter, I, therefore, proposed a hypothesis for a solution: what allows the synthesis between software and experience is an act of design, that is, an interpretation. We need to include a design factor in the software definition.

The first chapter opened the way followed by the second. My thesis is that there exists an extra-functional dimension in software, which is a hermeneutic dimension. Before being an algorithm, code, or problem solver, software is an act of interpretation. The analogy between software and text helped me to develop this point. Obviously, software is not text but something much more complex and multifaceted. However, the thesis of this book is that in software there is a hermeneutic process (the dialectics of distancing-belonging) very similar to that in the text, according to Ricoeur's view. Software re-writes all forms of technologically mediated human experience. I have tried to show how this conception is confirmed and enriched by the critical code studies.

The third chapter of the book is the most complex one. In this chapter, I hold two theses. The first: software is a process consisting of different levels and in the passage from one level to another a "regression" to the materiality of writing takes place. The computer does not see the meaning, but only the written traces in the silicon. The second: even the simplest written trace involves a hermeneutic process. I have tried to show this point by analyzing (a) the birth of modern mathematical writing from Cardan to Descartes and Leibniz and (b) the origin and history of software in the twentieth century. My goal was to show how software can be seen as the last chapter of a long history, that of writing.

The fourth chapter unifies the two parts of the book, the historical (Chap. 4) and the theoretical (Chap. 3), from a Kantian perspective. The central thesis is that software is a form of reflective judgment, namely, "digital reflective judgement." This transcendental approach allows us to overcome the limitations of an overly dualistic and over-intellectualized conception of software. The condition of possibility of this judgment is what I have called "graphic mind," that is some graphic structures that have their basis in the concept of a diagram. From this point of view, this book has intended to be an extension of the theses that have been already claimed by other authors (see Sack 2019; Frabetti 2015; Chun 2013).

The concept of digital reflective judgment is particularly important in my analysis because it completes above all the hermeneutic redefinition of the software that I gave in Chap. 3, following the Ricoeur model. Moreover, Ricoeur himself speaks of the text and the story as forms of reflective judgment (Ricoeur 1984, 32).

I do not think that defining software from a hermeneutic point of view is a way to undermine and weaken software as well as the rationality that lies at its root. I do not consider hermeneutics as the expression of a "weak" rationality as opposed to the "strong" rationality of STEM disciplines. I think that some ways of thinking about the concepts of hermeneutics and interpretation—in particular those inspired by Heidegger or that try to mimic Heidegger—are "weak" in themselves, in the sense that they represent a too metaphysical and self-referential form of philosophy at the borders with theology. From my point of view, hermeneutics represents a critique of the concept of pure, self-centered, self-sufficient, closed rationality. The hermeneutical point of view coincides with the thesis according to which any possible cognitive act always has presuppositions that are not explicit, or that cannot be explicit, because they are variables whose definition is not univocal—the definition of the variable either is not possible or is always different according to the reference context. What do I mean by this? By interpretation I do not mean a "way of being" or a certain type of "understanding," but two key aspects of human experience: (a) the dimension of meaning and (b) the absence of meaning and the need to create it. What is the meaning? The answers could be innumerable, and numerous books would have to be written to deal with the problem. Here I limit myself to advancing a hypothesis: the meaning is the emergent property (O'Connor 2020) of some physical states of the human brain. The interpretation designates the moment in which this property is absent, and its production is not immediate; there are microphysical states of the brain to which meanings do not correspond. Therefore, the meaning at the same time causally depends on certain states of the brain, but it is autonomous of them.

I do not want to go into the discussion now of what an emergent property is or how it can be understood. I limit myself to underlining that the concept of emergent property gives us a useful tool to overcome the brain-mind dualism and to understand the place of hermeneutics. As Michel (2019) points out, interpretation arises when the meaning (of a speech, of an action, etc.) is not clear or is not entirely there. The absence of meaning makes interpretation necessary. The absence of meaning is a condition that deeply characterizes the human being. The more the knowledge and experience of the world increase, the more complex it is to make sense of things. For this reason, before being an exegesis of the text, interpretation is, first of all, a creative activity, that is, the creation of meaning. And it is therefore everywhere in human activity, from action to speech, from the humanities to the exact sciences. The mere choice of how to use our terms and in which contexts to do so is interpretation. And therefore also the algorithm, as a problem solving, requires an interpretation, which is a basic choice on how to build meaning.

## The Critical Code Thinking

This book aimed to be a first step toward a larger project, what I call critical code thinking. I want to anticipate in this section some research lines of this project.

The main objective of critical code thinking is the study of software from a cultural and social perspective to identify what Feenberg (2010) called "formal biases," that is, cultural and social prejudices and habits embedded in a technological system. "Formal bias prevails wherever the structure or context of rationalized systems or institutions favors a particular social group" (Feenberg 2010, 163). In reformulating the Frankfurt school's Marxist approach, Feenberg distinguishes between two types of formal bias based on their relationship with the technological system: "Constitutive biases" are "values embodied in the nature or design of a theoretical system or artifact," whereas "implementation biases" are "values realized through contextualization" (163). One example of constitutive bias is surveillance systems, which "are biased by their very nature" because "their effect is to enhance the power of a minority at the expense of a majority, the surveilled" (164). Even a scientific theory can contain constitutive biases because "the constitution of an object of science depends on valuative decisions about epistemic methods" (164). One example of implementation bias could include city planning: "Urban plans that concentrate waste dumps near racial minorities are biased by the way in which the dumps relate to a context, not by the fact of their nature or design" (164).

An objector might say: the Marxian-inspired critics of software as a capitalist tool, borrowing from Castoriadis and the Frankfurt School is just a plea but not an argumentation. In fact—the objector would say—there will always be biases in software as narratives. Denouncing software as being biased, for example, by promoting some social (dominant) group, seems to suppose there can be software without biases if we pay attention close enough to this issue. But this is impossible.

This criticism is true. However, the intent of critical code thinking is not to eliminate biases from software, or to plan a political action against software companies, but to analyze these biases and understand, through them, what software can tell us about our society, the world, and the time we live in. If narratives are always ideological—even Ricoeur connects the concept of ideology to that of narration (see Ricoeur 2016)—then studying the ideology of software is a way of studying its being a narrative.

Therefore, critical code thinking's goal must be to develop a series of strategies to analyze the relationship between software and a capitalist society, or what Fisher has called "capitalist realism" (Fisher 2009), that is, the idea according to which post-Fordist liberal capitalism, together with its political system of reference (parliamentarism), is the only possible and acceptable reality, a natural and inevitable solution. As Fisher and Zizek have pointed out, capitalism harnesses the deepest roots of human desire to create an egalitarian force that flattens and assigns a monetary value to every object, belief, and symbol, as well as nullifying differences. Present, past, and future are reduced to a single flat, repetitive dimension. Perhaps no one besides Fisher has ever pointed out that today, even rebellion against capitalism—from alternative and independent cultural areas—has become the dominant style of capitalism itself. In post-Fordist capitalism, nothing works better than the criticism and condemnation of capitalism itself, the mechanism Fisher calls precorporation.

I, therefore, propose at the end of this book to integrate the Ricoeurian hermeneutical approach with Feenberg's theory of instrumentalization. For Feenberg—who was influenced profoundly by Heidegger, Marx, and Marcuse—technology is not an object, but rather a process of contextualization and recontextualization that involves two distinct levels: (a) primary instrumentalization, which concerns the function of the artifact connected to human intention, and (b) secondary instrumentalization, which concerns cultural and social meanings that the artifact receives from its use. To become technology properly, an object first must acquire a function and become de-contextualized, that is, removed from its original context and connected to a human teleology. The object is a certain amount of matter that is shaped according to a certain form of human intentionality. Thanks to the acquired function, the object is subsequently recontextualized in the sense that it becomes part of a new social context. During this second phase, the object acquires a series of cultural and social meanings or qualities that do not derive from its materiality or function. These qualities can be narrative (the object is connected to a narrative, e.g., to a nation's history), legal, sociological, psychological, aesthetic, ethical, and imaginary (the collective imagination). Interestingly, for Feenberg, secondary instrumentalization occurs through the design process, which is very close to what we said at the end of this book's first chapter: "Design codes are sometimes explicitly formulated in specifications or regulations. But often they are implicit in culture, training, and design, and need to be extracted by sociological analysis" (Feenberg 2010, 178).

However, the function cannot be separated from the meaning, and it always unfolds in a specific social context. Secondary instrumentalization gives the function social and cultural qualities that make it understandable and acceptable in a certain social context and cultural system. It is a very complex process, as Feenberg writes in an important passage:

> The affordances must be cast in a form acceptable to eventual users situated in a definite social context. Since technical workers usually share much of that context, many secondary instrumentalizations occur more or less unconsciously. Others are the result of using previously designed materials that embody the effects of earlier social interventions. Still others are dictated by laws and regulations or management decisions. Technical workers are of course aware that they are building a product for a specific user community, and to some degree they design in accordance with an amateur sociology of the user. This sociologizing task may be assigned to others in the organizations for which they work. (Feenberg 2010, 175)

Between primary and secondary instrumentalizations lie some intermediate levels, one of which, for Feenberg, is the "technical code," representing the convergence between social and technical needs: "A technical code describes the congruence of a social demand and a technical specification" (Feenberg 2010, 68). The "technical code" is a translation process between two different registers, one discursive and one technical, that is, the technical process is never unique. It implies a certain number of variables, that is, many possibilities to interpret the object's functions. The technical code simultaneously (a) restricts the number of variables in the process of interpreting the function and (b) restricts the concrete possibilities of satisfying them. Inevitably, the technical code is the expression of the dominant groups or hegemonic visions' values and needs: "Technical codes are always biased to some extent by the values of the dominant actors" (Feenberg 2010, 68). Technology tends to present itself as a neutral and extra-political dimension, with any social consequences from its functions deemed marginal effects from scientific and technological progress. However, Feenberg argues that, following Marcuse, social trends pervade technology and influence its development, and that secondary instrumentalization is much more powerful than primary instrumentalization.

The purpose of critical theory applied to technology must be to identify the limits of the technical codes at work in a certain technological process. Critical theory aims to create the best possible balance between first and second instrumentalizations, thereby intending to eliminate harmful biases and democratize technology by giving space to transparency, justice, respect for minorities (especially those excluded from the design process), and human rights. Thus, technology can be redesigned to serve humanity in the best possible way.

Can we apply these concepts to the study of software? I do not believe so. As I have tried to point out in this book, software is not simply technology, but rather redefines all technologies and social rationality itself. Today, software is the only true instrumentalization process. Contextualization and recontextualization always and only occur within the Stack—evoking Bratton's expression again. For example, some types of bias identifiable in the software industry are tied to the prevalence of males over females, the predominance of white males, or the acceptance and encouragement of certain psychological types or behaviors. An analysis of these biases should lead to a discussion about the relationship between social, economic, and technological development, for example, is there a necessary link between technological and economic development?

As explained at the beginning of this book, software undermines the unity of experience. Design is what creates a synthesis between software and human experience—what many authors would call *lifeworld*. Following Feenberg's indications, the task of critical code thinking should be not only to identify biases in software, but also to create new design solutions that can combine software and experience. This means not only finding new narrative forms but also new political solutions for the production and use of software. Following Feenberg, we could say that "the experienced lifeworld and the nature of natural science do not just coexist side by side," but rather, "they interact in many ways." Software is a third world that stands between the two others.

As Fisher (2009) writes, capitalism is characterized by the same contradiction that characterizes real socialism. On one hand, the official culture presents businesses and companies as caring and socially responsible. On the other hand, widespread awareness exists that these companies are actually ruthless and inhuman. One example is the paradox of the call center, an indifferent, impersonal, abstract, fragmentary system without a center and managed by poorly prepared and uninterested people, representing itself, nevertheless, through cheerful and joyful promotional campaigns. Late capitalism implements strategies through public relations, marketing, advertising, extreme forms of bureaucracy (target, mission, objectives, achievements, *nomeklatura*, the culture of auditing, the need to be "smart," etc.), and more complex ideologies or meta-narratives that aim to hide this contradiction. Capitalism, in the name of efficiency and the rejection of illusions, has built an immense, symbolic plot around itself that deforms reality in the name of reality itself. A system of constant deferral—a labyrinth in which no one knows what a mysterious ultimate authority is asking of them—is useless and demoralizing. The bureaucratic procedures become autonomous entities that float freely, without any restrictions, for the sole purpose of maximizing anxiety and ensuring that individuals introject the control apparatus as much as possible: "The invocation of the idea that 'there is no alternative' and the recommendation to 'work smarter, not harder' shows how capitalist realism sets the tone for labor disputes in post-Fordism" (Fisher 2009, 57).

Software is the perfect tool that capitalism has produced to mask its internal contradictions and withdraw into invisibility. As Jameson pointed out in *The Antinomies of Realism* (2013), due to the decline of criticism and an alternative political vision, late twentieth-century capitalism profoundly transformed the sense of history and time. Dominating the horizon of humans living in this form of capitalism is a purely fungible present in which both the psyche and body can be quantified, broken down, and reconstructed as needed. To sum up, as Fisher (2009, 59) also states, late-capitalist reality becomes a digital document that is perpetually modifiable, but for this reason, also perennially unstable and fragile: "In conditions where realities and identities are upgraded like software, it is not surprising that memory disorders should have become the focus of cultural anxiety—see, for instance, the *Bourne* films, *Memento, Eternal Sunshine of the Spotless Mind*" (Fisher 2009, 62). The Stack is the new "face" of capitalism, comprising ubiquitous, elusive, intangible, and incomprehensible processes. The Stack is the new Tower of Babel of that *negative atheology* that Fisher (2009, 69) places at the center of capitalism: "The center is missing, but we cannot stop searching for it or positing it. It is not that there is nothing there—it is that what is there is not capable of exercising responsibility."

In this context, it is important to cite Dobson's thesis because it allows us to overturn the perspective. Late capitalism exploited and exploits software's potential, and yet, there is also a perverse effect: "Software is not only eating the world, as Marc Andreessen said, but it is also eating the capitalism itself—from the inside out." Dobson describes a vicious cycle: Capitalism, thanks to digital technologies and automation, has produced a total disconnect between productivity and work, in the sense that less and less work—and, therefore, workers—is required to produce the goods and services that human society needs. Capital's tendency is to reduce the cost of labor to the maximum and increase productivity, but this will have the effect of creating masses of unemployed people without any income, so they no longer will be able to consume, that is, take advantage of these goods and services that capital produces. In other words, thanks to software, capital has become more and more ontologically and geographically ubiquitous, but precisely software is consuming it: "Software contributes to this process. Which is why I say it is eating capitalism."[1] According to Dobson, software has put an end to one phase of capitalism and opened the doors to an entirely new post-capitalist society based on sharing goods, as demonstrated by the phenomenon of open sources. Therefore, software represents the most powerful, but also the most dangerous, tool for capitalism.

In conclusion, the study and analysis of software represent a crucial aspect to understand the reality in which the human beings of the twenty-first century live. Software looks like the new Golem, as in Gustav

Meyrink's novel. In the Jewish tradition, the Golem is a clay mannequin artificially shaped by humans through some ritual formulas, that is, a set of statement. In one version of the legend, the mannequin comes to life directly from writing, that is, by writing a single word on its forehead, or by inserting a parchment in his mouth. In Meyrink's novel, on the other hand, the Golem is something much more elusive and enigmatic in a mysterious Prague. The Golem is both a gift from God and a threat. It is both an opportunity and a risk. It is a border between two realities, the human and the divine. It is therefore in itself a paradoxical, restless being, without a precise identity, but which exercises enormous power.

## Note

1. https://thenewstack.io/oped-software-not-just-eating-world-capitalism/.

## References

Chun, W. 2013. *Programmed Visions. Software and Memory*. Cambridge, MA: MIT Press.
Feenberg, A. 2010. *Between Reason and Experience*. Cambridge, MA: MIT Press.
Fisher, M. 2009. *Capitalist Realism: Is There No Alternative?* London: Zero Books.
Frabetti, F. 2015. *Software Theory*. London & New York: Media Philosophy.
Jameson, F. 2013. *The Antinomies of Realism*. New York: Verso Books.
Marino, M. 2020. Critical Code Studies. Cambridge, MA: MIT Press.
Michel, J. 2019. *Homo Interpretans. Towards a Transformation of Hermeneutics*. London: Rowman & Littlefield.
O'Connor, T. 2020. Emergent Properties. *Stanford Encyclopedia of Philosophy*.
Possati, L., and Romele, A. 2021. Digital Hermeneutics. Introduction. Special issue of *Critical Hermeneutics*.
Ricoeur, P. 1984. *Time and Narrative 1*. University of Chicago Press.
———. 2016. *Idéologie et utopie*. Paris: Seuil.
Romele, A. 2019. *Digital Hermeneutics*. London: Routledge.
Sack, W. 2019. *The Software Arts*. Cambridge, MA: MIT Press.

# Index[1]



[1] Note: Page numbers followed by 'n' refer to notes.



L. M. Possati, *Software as Hermeneutics*,
https://doi.org/10.1007/978-3-030-63610-4

FSC
www.fsc.org
MIX
Papier | Fördert
gute Waldnutzung
FSC® C083411

Kolibri 360 GmbH
Ferdinand-Jühlke-Straße 7
99095 Erfurt, Deutschland
produktsicherheit@kolibri360.de